Studies in Systems, Decision and Control

Volume 411

The series "Studies in Systems, Decision and Control" (SSDC) covers both new developments and advances, as well as the state of the art, in the various areas of broadly perceived systems, decision making and control–quickly, up to date and with a high quality. The intent is to cover the theory, applications, and perspectives on the state of the art and future developments relevant to systems, decision making, control, complex processes and related areas, as embedded in the fields of engineering, computer science, physics, economics, social and life sciences, as well as the paradigms and methodologies behind them. The series contains monographs, textbooks, lecture notes and edited volumes in systems, decision making and control spanning the areas of Cyber-Physical Systems, Autonomous Systems, Sensor Networks, Control Systems, Energy Systems, Automotive Systems, Biological Systems, Vehicular Networking and Connected Vehicles, Aerospace Systems, Automation, Manufacturing, Smart Grids, Nonlinear Systems, Power Systems, Robotics, Social Systems, Economic Systems and other. Of particular value to both the contributors and the readership are the short publication timeframe and the world-wide distribution and exposure which enable both a wide and rapid dissemination of research output.

Indexed by SCOPUS, DBLP, WTI Frankfurt eG, zbMATH, SCImago.

All books published in the series are submitted for consideration in Web of Science.

More information about this series at https://link.springer.com/bookseries/13304

George A. Anastassiou

Abstract Fractional Monotone Approximation, Theory and Applications

Springer

George A. Anastassiou
Department of Mathematical Sciences
University of Memphis
Memphis, TN, USA

ISSN 2198-4182 ISSN 2198-4190 (electronic)
Studies in Systems, Decision and Control
ISBN 978-3-030-95942-5 ISBN 978-3-030-95943-2 (eBook)
https://doi.org/10.1007/978-3-030-95943-2

This Springer imprint is published by the registered company Springer Nature Switzerland AG
The registered company address is: Gewerbestrasse 11, 6330 Cham, Switzerland

Dedicated to: "Hard working unknown mathematician"

Preface

Fractional calculus and its long history started in 1695 by L'Hospital and Leibnitz. Over the years, many different kinds of it emerged according to the needs of science given by many famous mathematicians. The last 50 years fractional calculus due to its wide applications to many applied sciences has become very popular.

Its predominant kind is the newer one of Caputo type with all of its variations because of their easier applicability. These involve usually singular kernels but recently also non-singular kernels.

Almost all fractional calculi engage memory in their kernels and the two divisions of singular and non-singular kernel have their own advantages and disadvantages, but both have their strong merits.

In this short monograph, we employ an abstract kernel fractional calculus with applications to Prabhakar and non-singular kernel fractional calculi. Our results are univariate and bivariate. In the univariate case, we present abstract fractional monotone approximation by polynomials and splines, and in the bivariate case, we give the abstract fractional monotone constrained approximation by bivariate pseudo-polynomials and polynomials.

So in this monograph, all presented is original work by the author given at a very abstract level to cover a maximum number of different fractional calculi applications.

As a result, this short monograph is the natural and expected evolution of recent author's research work put in a book form for the first time. The presented approaches are original, and chapters are self-contained and can be read independently. This monograph is suitable to be used in related graduate classes and research projects.

The motivation to write this monograph came by the following: Various issues related to the modeling and analysis of ordinary and fractional-order systems have gained an increased popularity, as witnessed by many books and volumes in Springer's program: http://www.springer.com/gp/search?query=fractional&submit=Prze%C5%9Blij, and the purpose of our book is to capture at a very general level a deeper formal analysis on some issues that are relevant to many areas for instance: decision making, complex processes, systems modeling and control, and related areas.

The above are deeply embedded in the fields of mathematics, engineering, computer science, physics, economics, social, and life sciences.

Next listed are the author's recent monographs in fractional analysis and applications:

- George Anastassiou, "Intelligent Computations: Abstract Fractional Calculus, Inequalities, Approximations", Springer, Studies in Computational Intelligence 734, Heidelberg, New York, 2018.
- George Anastassiou, "Nonlinearity: Ordinary and Fractional Approximations by Sublinear and Max-product Operators", Springer, Studies in Systems, Decision and Control 147, Heidelberg, New York, 2018.
- George Anastassiou, "Ordinary and Fractional approximation by non-additive integrals: Choquet, Shilkret and Sugeno integral approximators", Springer, Studies in Systems, Decision and Control 190, Heidelberg, New York, 2019.
- George Anastassiou, "Intelligent Analysis: Fractional Inequalities and Approximations Expanded", Springer, Studies in Computational Intelligence 886, Heidelberg, New York, 2020,
- George Anastassiou, "Generalized Fractional Calculus: New Advancements and Applications", Springer, Studies in Systems, Decision and Control 305, Heidelberg, New York, 2021.
- George Anastassiou, "Constructive Fractional Analysis with Applications", Springer, Studies in Systems, Decision and Control 362, Heidelberg, New York, 2021.
- George Anastassiou, "Unification of Fractional Calculi with Applications", Springer, Studies in Systems, Decision and Control 398, Heidelberg, New York, 2022.

The complete list of presented topics is as follows:

- basic abstract fractional monotone polynomial approximation with applications,
- univariate simultaneous high-order abstract fractional monotone polynomial approximation with applications,
- simultaneous high-order abstract fractional monotone spline approximation and applications,
- abstract bivariate left fractional pseudo-polynomial monotone constrained approximation with applications,
- abstract bivariate right fractional pseudo-polynomial monotone constrained approximation and applications, and
- bivariate abstract fractional monotone constrained approximation by polynomials.

The book's results are expected to find applications in many areas of pure and applied mathematics, especially in fractional approximation theory and fractional differential equations. Other possible applications can be in applied sciences like geophysics, physics, chemistry, economics, and engineering, etc. All in all what is presented here is a valuable tool for a large range of applications. Therefore, this short monograph is suitable for researchers, graduate students, practitioners, and seminars of the above disciplines, also to be in all science and engineering libraries.

The preparation of the book took place during 2021 at the University of Memphis, at stay home of the author during the COVID-19 outbreak keeping him alive, happy, and in control!

The author likes to thank Prof. Alina Lupas of the University of Oradea, Romania, for checking and reading the manuscript.

Memphis, TN, USA
December 2021

George A. Anastassiou

Contents

Chapter 1
Basic Abstract Fractional Monotone Approximation

Here we extend our earlier fractional monotone approximation theory to abstract fractional monotone approximation, with applications to Prabhakar fractional calculus and non-singular kernel fractional calculi. We cover both the left and right sides of this constrained approximation. Let $f \in C^p([-1,1])$, $p \geq 0$ and let L be a linear abstract left or right fractional differential operator such that $L(f) \geq 0$ over $[0,1]$ or $[-1,0]$, respectively. We can find a sequence of polynomials Q_n of degree $\leq n$ such that $L(Q_n) \geq 0$ over $[0,1]$ or $[-1,0]$, respectively. Additionally f is approximated quantitatively with rates uniformly by Q_n with the use of first modulus of continuity of $f^{(p)}$. It follows [5].

1.1 Introduction

The topic of monotone approximation initiated in 1965, [15] by O. Shisha, and became a major trend in approximation theory. The original problem was: given a positive integer k, approximate with rates a given function f whose kth derivative is ≥ 0 by polynomials $(Q_n)_{n\in\mathbb{N}}$ having the same property.

In 1985, [6], the author and O. Shisha, continued this study by replacing the kth derivative with a linear differential operator of order k involving ordinary derivatives, again the approximation was with rates.

Later in 1991, [1], the author extended this kind of study in two dimensions, etc.

In 2015, [1], see Chaps. 1–8, the author went a step further, by starting the fractional monotone approximation. In that the linear differential operator is a fractional one, involving left or right side Caputo fractional derivatives.

To give a flavor of it we need:

Definition 1.1 ([8], p. 50) Let $\alpha > 0$ and $\lceil \alpha \rceil = m \in \mathbb{N}$ ($\lceil \cdot \rceil$ is the ceiling of the number). Consider $f \in C^m([-1,1])$. We define the left side Caputo fractional derivative

G. A. Anastassiou, *Abstract Fractional Monotone Approximation, Theory and Applications*, Studies in Systems, Decision and Control 411,
https://doi.org/10.1007/978-3-030-95943-2_1

of f of order α as follows:

$$\left(D_{*-1}^{\alpha} f\right)(x)=\frac{1}{\Gamma(m-\alpha)} \int_{-1}^{x}(x-t)^{m-\alpha-1} f^{(m)}(t)\, dt, \tag{1.1}$$

for any $x \in [-1, 1]$, where Γ is the gamma function. We set

$$D_{*-1}^{0} f(x)=f(x), \ D_{*-1}^{m} f(x)=f^{(m)}(x), \ \forall x \in[-1,1].$$

Also to motivate our work we mention:

Theorem 1.2 *([2], p. 2) Let h, k, p be integers, $0 \leq h \leq k \leq p$ and let f be a real function, with $f^{(p)}$ continuous in $[-1, 1]$ and first modulus of continuity $\omega_1\left(f^{(p)}, \delta\right)$, where $\delta > 0$. Let $\alpha_j(x)$, $j = h, h+1, ..., k$ be real functions, defined and bounded on $[-1, 1]$ and assume for $x \in [0, 1]$ that $\alpha_h(x)$ is either $\geq$ some number $\alpha > 0$ or $\leq$ some number $\beta < 0$. Let the real numbers $\alpha_0 = 0 < \alpha_1 \leq 1 < \alpha_2 \leq 2 < ... < \alpha_p \leq p$. Here $D_{*-1}^{\alpha_j} f$ stands for the left Caputo fractional derivative of f of order α_j anchored at -1. Consider the linear left fractional differential operator*

$$L := \sum_{j=h}^{k} \alpha_j(x)\left[D_{*-1}^{\alpha_j}\right] \tag{1.2}$$

and suppose, throughout $[0, 1]$,

$$L(f) \geq 0. \tag{1.3}$$

Then, for any $n \in \mathbb{N}$, there exists a real polynomial $Q_n(x)$ of degree $\leq n$ such that

$$L(Q_n) \geq 0 \text{ throughout } [0,1], \tag{1.4}$$

and

$$\max_{-1 \leq x \leq 1}\left|f(x)-Q_n(x)\right| \leq C n^{k-p} \omega_1\left(f^{(p)}, \frac{1}{n}\right), \tag{1.5}$$

where C is a constant independent of n or f.

As you see the monotonicity property here is true only on the critical interval $[0, 1]$.

We will use the following important result:

Theorem 1.3 *(see: [17] by S.A. Teljakovskii and [18] by R.M. Trigub) Let $n \in \mathbb{N}$. Be given a real function g, with $g^{(p)}$ continuous in $[-1, 1]$, there exists a real polynomial $q_n(x)$ of degree $\leq n$ such that*

$$\max_{-1 \leq x \leq 1}\left|g^{(j)}(x)-q_n^{(j)}(x)\right| \leq R_p n^{j-p} \omega_1\left(g^{(p)}, \frac{1}{n}\right), \tag{1.6}$$

$j = 0, 1, \ldots, p$, *where* R_p *is a constant independent of* n *or* g.

In this chapter we perform abstract fractional calculus, left and right monotone approximation theory of Caputo type, and then we apply our results to Prabhakar fractional Calculus, generalized non-singular fractional calculus, and parametrized Caputo-Fabrizio non-singular fractional calculus.

Next, we build the related necessary fractional calculi background.

1.2 Fractional Calculi

Here we need to be very specific in preparation for our main results.

1.2.1 Abstract Fractional Calculus

Let $h, k \in \mathbb{Z}_+$, $p \in \mathbb{N} : 0 \leq h \leq k \leq p$. Let also $\mathbb{N} \not\ni \alpha_j > 0$, $j = 1, \ldots, p$, such that $\alpha_0 = 0 < \alpha_1 < 1 < \alpha_2 < 2 < \alpha_3 < 3 < \ldots < \ldots < \alpha_p < p$. That is $\lceil \alpha_j \rceil = j$, $j = 1, \ldots, p$; $\lceil \alpha_0 \rceil = 0$.

Consider the integrable functions $k_j := K_{\alpha_j} : [0, 2] \to \mathbb{R}_+$, $j = 0, 1, \ldots, p$. Here $g \in C^p([-1, 1])$.

We consider the following abstract left side Caputo type fractional derivatives:

$$\left({}^{k_j}D_{*-1}^{\alpha_j} g\right)(x) := \int_{-1}^{x} k_j(x - t)\, g^{(j)}(t)\, dt, \tag{1.7}$$

$j = 1, \ldots, p$; $\forall\, x \in [-1, 1]$.

Similarly, we define the corresponding right side generalized Caputo type fractional derivatives:

$$\left({}^{k_j}D_{1-}^{\alpha_j} g\right)(x) := (-1)^j \int_{x}^{1} k_j(t - x)\, g^{(j)}(t)\, dt, \tag{1.8}$$

$j = 1, \ldots, p$; $\forall\, x \in [-1, 1]$.

We set

$$\left({}^{k_j}D_{*-1}^{j} g\right)(x) := g^{(j)}(x); \quad \left({}^{k_j}D_{1-}^{j} g\right)(x) := (-1)^j g^{(j)}(x), \tag{1.9}$$

for $j = 1, \ldots, p$, and also we set

$$\left({}^{k_0}D_{*-1}^{0} g\right)(x) := \left({}^{k_0}D_{1-}^{0} g\right)(x) := g(x), \tag{1.10}$$

$\forall x \in [-1, 1]$.

We will assume that

$$\int_0^1 k_h(z)\, dz \geq 1, \text{ when } h \neq 0. \tag{1.11}$$

In the usual Caputo fractional derivatives case it is

$$k_j(z) = \frac{z^{j-\alpha_j-1}}{\Gamma(j-\alpha_j)}, \quad j = 1, ..., p; \forall z \in [0, 2], \tag{1.12}$$

and (1.11) is fulfilled, by the fact that $\Gamma(h - \alpha_h + 1) \leq 1$, see [2], p. 6.

1.2.2 About Prabhakar Fractional Calculus

Here we follow [3, 13].

We consider the Prabhakar function (also known as the three parameter Mittag-Leffler function), (see [10], p. 97; [9])

$$E_{\alpha,\beta}^{\gamma}(z) = \sum_{k=0}^{\infty} \frac{(\gamma)_k}{k!\Gamma(\alpha k + \beta)} z^k, \tag{1.13}$$

where Γ is the gamma function; $\alpha, \beta > 0$, $\gamma \in \mathbb{R}$, $z \in \mathbb{R}$, and $(\gamma)_k = \gamma(\gamma+1)\ldots(\gamma+k-1)$. It is $E_{\alpha,\beta}^{0}(z) = \frac{1}{\Gamma(\beta)}$.

Let $a, b \in \mathbb{R}$, $a < b$ and $x \in [a, b]$; $f \in C([a, b])$. The left and right Prabhakar fractional integrals are defined ([3, 13]) as follows:

$$\left(e_{\rho,\mu,\omega,a+}^{\gamma} f\right)(x) = \int_a^x (x-t)^{\mu-1} E_{\rho,\mu}^{\gamma}\left[\omega(x-t)^{\rho}\right] f(t)\, dt, \tag{1.14}$$

and

$$\left(e_{\rho,\mu,\omega,b-}^{\gamma} f\right)(x) = \int_x^b (t-x)^{\mu-1} E_{\rho,\mu}^{\gamma}\left[\omega(t-x)^{\rho}\right] f(t)\, dt, \tag{1.15}$$

where $\rho, \mu > 0$; $\gamma, \omega \in \mathbb{R}$.

Functions (1.14) and (1.15) are continuous, see [3].

Next, let $f \in C^N([a, b])$, where $N = \lceil \mu \rceil$, ($\lceil \cdot \rceil$ is the ceiling of the number), $0 < \mu \notin \mathbb{N}$. We define the Prabhakar-Caputo left and right fractional derivatives of order μ ([3, 13]) as follows ($x \in [a, b]$):

$$\left({}^{C}D^{\gamma}_{\rho,\mu,\omega,a+}f\right)(x) = \int_{a}^{x} (x-t)^{N-\mu-1} E^{-\gamma}_{\rho,N-\mu}\left[\omega (x-t)^{\rho}\right] f^{(N)}(t)\,dt, \tag{1.16}$$

with ${}^{C}D^{\gamma}_{\rho,0,\omega,a+}f := f$; ${}^{C}D^{\gamma}_{\rho,N,\omega,a+}f := f^{(N)}$, $N \in \mathbb{N}$,
and

$$\left({}^{C}D^{\gamma}_{\rho,\mu,\omega,b-}f\right)(x) = (-1)^{N} \int_{x}^{b} (t-x)^{N-\mu-1} E^{-\gamma}_{\rho,N-\mu}\left[\omega (t-x)^{\rho}\right] f^{(N)}(t)\,dt, \tag{1.17}$$

with ${}^{C}D^{\gamma}_{\rho,0,\omega,b-}f := f$; ${}^{C}D^{\gamma}_{\rho,N,\omega,b-}f := (-1)^{N} f^{(N)}$, $N \in \mathbb{N}$.

One can rewrite (1.16) and (1.17) as

$$\left({}^{C}D^{\gamma}_{\rho,\mu,\omega,a+}f\right)(x) = \left(e^{-\gamma}_{\rho,N-\mu,\omega,a+}f^{[N]}\right)(x), \tag{1.18}$$

and

$$\left({}^{C}D^{\gamma}_{\rho,\mu,\omega,b-}f\right)(x) = (-1)^{N}\left(e^{-\gamma}_{\rho,N-\mu,\omega,b-}f^{[N]}\right)(x), \tag{1.19}$$

$\forall\, x \in [a, b]$.

Clearly, the functions (1.18) and (1.19) are continuous.

By [3], we have that

$$\left(e^{\gamma}_{\rho,\mu,\omega,a+}f\right)(x) = \int_{a}^{x} (x-t)^{\mu-1} \left(\sum_{k=0}^{\infty} \frac{(\gamma)_k}{k!\Gamma(\rho k+\mu)} \left(\omega (x-t)^{\rho}\right)^{k}\right) f(t)\,dt$$

$$= \sum_{k=0}^{\infty} \frac{(\gamma)_k\, \omega^k}{k!\Gamma(\rho k+\mu)} \int_{a}^{x} (x-t)^{(\rho k+\mu)-1} f(t)\,dt, \tag{1.20}$$

$\forall\, x \in [a, b]$.

That is

$$\left(e^{\gamma}_{\rho,\mu,\omega,a+}1\right)(x) = \sum_{k=0}^{\infty} \frac{(\gamma)_k\, \omega^k}{k!\Gamma(\rho k+\mu)} \int_{a}^{x} (x-t)^{(\rho k+\mu)-1}\,dt =$$

$$\sum_{k=0}^{\infty} \frac{(\gamma)_k\, \omega^k}{k!\Gamma(\rho k+\mu)} \frac{(x-a)^{(\rho k+\mu)}}{(\rho k+\mu)} = \sum_{k=0}^{\infty} \frac{(\gamma)_k\, \omega^k}{k!\Gamma(\rho k+\mu+1)} (x-a)^{(\rho k+\mu)} = \tag{1.21}$$

$$(x-a)^{\mu} \sum_{k=0}^{\infty} \frac{(\gamma)_k}{k!\Gamma(\rho k+\mu+1)} \left[\omega (x-a)^{\rho}\right]^{k} = (x-a)^{\mu} E^{\gamma}_{\rho,\mu+1}\left(\omega (x-a)^{\rho}\right).$$

So we have proved that

$$\left(e^{\gamma}_{\rho,\mu,\omega,a+}1\right)(x) = (x-a)^{\mu} E^{\gamma}_{\rho,\mu+1}\left(\omega (x-a)^{\rho}\right), \tag{1.22}$$

$\forall\, x \in [a,b]$.

Similarly, we have

$$\left(e^{\gamma}_{\rho,\mu,\omega,b-}1\right)(x) = \sum_{k=0}^{\infty} \frac{(\gamma)_k\, \omega^k}{k!\Gamma(\rho k+\mu)} \int_x^b (t-x)^{(\rho k+\mu)-1}\, dt =$$

$$\sum_{k=0}^{\infty} \frac{(\gamma)_k\, \omega^k}{k!\Gamma(\rho k+\mu)} \frac{(b-x)^{(\rho k+\mu)}}{(\rho k+\mu)} = \sum_{k=0}^{\infty} \frac{(\gamma)_k\, \omega^k}{k!\Gamma(\rho k+\mu+1)} (b-x)^{(\rho k+\mu)} = \tag{1.23}$$

$$(b-x)^{\mu} \sum_{k=0}^{\infty} \frac{(\gamma)_k}{k!\Gamma(\rho k+\mu+1)} \left[\omega (b-x)^{\rho}\right]^k = (b-x)^{\mu} E^{\gamma}_{\rho,\mu+1}\left(\omega (b-x)^{\rho}\right).$$

So it holds

$$\left(e^{\gamma}_{\rho,\mu,\omega,b-}1\right)(x) = (b-x)^{\mu} E^{\gamma}_{\rho,\mu+1}\left(\omega (b-x)^{\rho}\right), \tag{1.24}$$

$\forall\, x \in [a,b]$.

Next we take $[a,b] = [-1,1]$.

Thus

$$\left(e^{\gamma}_{\rho,\mu,\omega,-1+}1\right)(x) = (x+1)^{\mu} E^{\gamma}_{\rho,\mu+1}\left(\omega (x+1)^{\rho}\right), \tag{1.25}$$

and

$$\left(e^{\gamma}_{\rho,\mu,\omega,1-}1\right)(x) = (1-x)^{\mu} E^{\gamma}_{\rho,\mu+1}\left(\omega (1-x)^{\rho}\right), \tag{1.26}$$

$\forall\, x \in [-1,1]$.

Clearly, then we get

$$\left(e^{-\gamma}_{\rho,N-\mu,\omega,-1+}1\right)(x) = (x+1)^{N-\mu} E^{-\gamma}_{\rho,N-\mu+1}\left(\omega (x+1)^{\rho}\right), \tag{1.27}$$

and

$$\left(e^{-\gamma}_{\rho,N-\mu,\omega,1-}1\right)(x) = (1-x)^{N-\mu} E^{-\gamma}_{\rho,N-\mu+1}\left(\omega (1-x)^{\rho}\right), \tag{1.28}$$

$\forall\, x \in [-1,1]$.

Here it is $N-\mu > 0$. By assumption we take $\rho > 0$, $\gamma < 0$ and for convenience we consider only $\omega > 0$.

Therefore, we derive the basic Hardy type inequalities:

$$\left\| e^{-\gamma}_{\rho,N-\mu,\omega,-1+}1 \right\|_{\infty,[-1,1]} \leq 2^{N-\mu} E^{-\gamma}_{\rho,N-\mu+1}\left(2^{\rho}\omega\right), \tag{1.29}$$

and

$$\left\| e^{-\gamma}_{\rho, N-\mu, \omega, 1-} 1 \right\|_{\infty, [-1,1]} \leq 2^{N-\mu} E^{-\gamma}_{\rho, N-\mu+1} \left(2^{\rho} \omega\right). \tag{1.30}$$

1.2.3 From Generalized Non-singular Fractional Calculus ([4])

We need

Definition 1.4 Here we use the multivariate analogue of generalized Mittag-Leffler function, see [14], defined for $\lambda, \gamma_j, \rho_j, z_j \in \mathbb{C}$, $Re\left(\rho_j\right) > 0\,(j = 1, ..., m)$ in terms of a multiple series of the form:

$$E^{(\gamma_j)}_{(\rho_j),\lambda}(z_1, ..., z_m) = E^{(\gamma_1,...,\gamma_m)}_{(\rho_1,...,\rho_m),\lambda}(z_1, ..., z_m) =$$

$$\sum_{k_1,...,k_m=0}^{\infty} \frac{(\gamma_1)_{k_1} \cdots (\gamma_m)_{k_m}}{\Gamma\left(\lambda + \sum_{j=1}^{m} k_j \rho_j\right)} \frac{z_1^{k_1} ... z_m^{k_m}}{k_1! ... k_m!}, \tag{1.31}$$

where $\left(\gamma_j\right)_{k_j}$ is the Pochhammer symbol, Γ is the gamma function. By [16], p. 157, (1.31) converges for $Re\left(\rho_j\right) > 0$, $j = 1, ..., m$.

In what follows we will use the particular case of $E^{(\gamma_1,...,\gamma_m)}_{(\rho,...,\rho),\lambda}\left[\omega_1 t^{\rho}, ..., \omega_m t^{\rho}\right]$, denoted by $E^{(\gamma_j)}_{(\rho),\lambda}\left[\omega_1 t^{\rho}, ..., \omega_m t^{\rho}\right]$, where $0 < \rho < 1$, $t \geq 0$, $\lambda > 0$, $\gamma_j \in \mathbb{R}$ with $\left(\gamma_j\right)_{k_j} := \gamma_j\left(\gamma_j + 1\right) ... \left(\gamma_j + k_j - 1\right)$, $\omega_j \in \mathbb{R} - \{0\}$, for $j = 1, ..., m$.

Let now $f \in C^{n+1}([a, b])$, $n \in \mathbb{Z}_+$.

We define the Caputo type generalized left fractional derivative with non-singular kernel of order $n + \rho$, as

$$D^{n+\rho}_{a*} f(x) := {}^{CA}_{(\gamma_j)(\omega_j)} D^{n+\rho,\lambda}_{a*} f(x) :=$$

$$\frac{A(\rho)}{1-\rho} \int_a^x E^{(\gamma_j)}_{(\rho),\lambda}\left[\frac{-\omega_1 \rho}{1-\rho}(x-t)^{\rho}, ..., \frac{-\omega_m \rho}{1-\rho}(x-t)^{\rho}\right] f^{(n+1)}(t)\, dt, \tag{1.32}$$

$\forall\, x \in [a, b]$.

Similarly, we define the Caputo type generalized right fractional derivative with non-singular kernel of order $n + \rho$, as

$$D_{b-}^{n+\rho} f(x) := {}_{(\gamma_j)(\omega_j)}^{CA} D_{b-}^{n+\rho,\lambda} f(x) :=$$

$$(-1)^{n+1} \frac{A(\rho)}{1-\rho} \int_x^b E_{(\rho),\lambda}^{(\gamma_j)} \left[\frac{-\omega_1 \rho}{1-\rho} (t-x)^{\rho}, ..., \frac{-\omega_m \rho}{1-\rho} (t-x)^{\rho} \right] f^{(n+1)}(t) \, dt, \tag{1.33}$$

$\forall x \in [a, b]$.

Above $A(\rho)$ is a normalizing constant.

The above derivatives (1.32), (1.33) generalize the Atangana-Baleanu fractional derivatives [7].

We mention the following Hardy type inequalities:

Theorem 1.5 *([4]) All as above with $\gamma_j > 0$, $j = 1, ..., m$; $\lambda = 1$. Then*

$$\left\{ \left\| D_{a*}^{n+\rho} f \right\|_{\infty}, \left\| D_{b-}^{n+\rho} f \right\|_{\infty} \right\} \leq \frac{(b-a) |A(\rho)|}{1-\rho}$$

$$E_{(\rho),2}^{(\gamma_j)} \left[\frac{|\omega_1| \rho}{1-\rho} (b-a)^{\rho}, ..., \frac{|\omega_m| \rho}{1-\rho} (b-a)^{\rho} \right] \left\| f^{(n+1)} \right\|_{\infty} < \infty, \tag{1.34}$$

where $n \in \mathbb{Z}_+$.

We also mention

Theorem 1.6 *([4]) All as above with $\gamma_j > 0$, $j = 1, ..., m$, and $\lambda > 0$, $0 < \rho < 1$, etc. Then*

$$D_{a*}^{n+\rho} f, \ D_{b-}^{n+\rho} f \in C([a, b]), \ n \in \mathbb{Z}_+.$$

We rewrite (1.32) and (1.33), and for $[a, b] = [-1, 1]$.

Let $\mu > 0$ with $\mu \notin \mathbb{N}$ and $\lceil \mu \rceil = n \in \mathbb{N}$. That is $0 < 1 - n + \mu < 1$, and let $f \in C^n([-1, 1])$. Then, we have

$$D_{-1*}^{\mu} f(x) := {}_{(\gamma_j)(\omega_j)}^{CA} D_{-1*}^{\mu,\lambda} f(x) :=$$

$$\frac{A(1-n+\mu)}{n-\mu} \int_{-1}^{x} E_{(1-n+\mu),\lambda}^{(\gamma_j)} \left[\frac{-\omega_1 (1-n+\mu)}{n-\mu} (x-t)^{1-n+\mu}, ..., \right. \tag{1.35}$$

$$\left. \frac{-\omega_m (1-n+\mu)}{n-\mu} (x-t)^{1-n+\mu} \right] f^{(n)}(t) \, dt,$$

and

$$D_{1-}^{\mu} f(x) := {}_{(\gamma_j)(\omega_j)}^{CA} D_{1-}^{\mu,\lambda} f(x) :=$$

$$(-1)^n \frac{A(1-n+\mu)}{n-\mu} \int_x^1 E^{(\gamma_j)}_{(1-n+\mu),\lambda} \left[\frac{-\omega_1 (1-n+\mu)}{n-\mu} (t-x)^{1-n+\mu}, \ldots, \right. \tag{1.36}$$

$$\left. \frac{-\omega_m (1-n+\mu)}{n-\mu} (t-x)^{1-n+\mu} \right] f^{(n)}(t)\, dt,$$

$\forall\, x \in [-1, 1]$.

We will set $D^0_{-1*} f = f$, $D^0_{1-} f = f$, and $D^m_{-1*} f = f^{(m)}$, $D^m_{1-} f = (-1)^m f^{(m)}$, when $m \in \mathbb{N}$.

We make

Remark 1.7 Fractional Calculi of Sects. 1.2.2 and 1.2.3 are special cases of abstract fractional calculus, see Sect. 1.2.1. In particular the important condition (1.11) is fulfilled.

So, we will verify $\int_0^1 k_h(z)\, dz \geq 1$, $h \neq 0$.

(I) First for Sect. 1.2.2:

We notice that

$$\int_0^1 z^{N-\mu-1} E^{-\gamma}_{\rho,N-\mu}(\omega z^\rho)\, dz =$$

(here $\rho, N-\mu > 0$, $\gamma < 0$, $\omega > 0$)

$$\int_0^1 z^{N-\mu-1} \sum_{k=0}^{\infty} \frac{(-\gamma)_k}{k!\Gamma(\rho k + N - \mu)} (\omega z^\rho)^k\, dz \overset{\text{(by [11], p. 175)}}{=}$$

$$\sum_{k=0}^{\infty} \frac{(-\gamma)_k \omega^k}{k!\Gamma(\rho k + N - \mu)} \int_0^1 z^{N-\mu-1} z^{\rho k} dz = \tag{1.37}$$

$$\sum_{k=0}^{\infty} \frac{(-\gamma)_k \omega^k}{k!\Gamma(\rho k + N - \mu)} \int_0^1 z^{(\rho k+N)-\mu-1} dz =$$

$$\sum_{k=0}^{\infty} \frac{(-\gamma)_k \omega^k}{k!\Gamma(\rho k + N - \mu + 1)} = E^{-\gamma}_{\rho,N-\mu+1}(\omega) \geq 1,$$

for suitable $\omega > 0$.

(II) Next, for Sect. 1.2.3:

Here $\gamma_j > 0$, j=1, ..., m; $\lambda = 1$; $\mathbb{N} \not\ni \mu > 0$, $\lceil \mu \rceil = n \in \mathbb{N}$, $\omega_j < 0$, $j = 1, \ldots, m$. Without loss of generality we assume that $A(1-n+\mu) > 0$.

We have that

$$\frac{A(1-n+\mu)}{n-\mu}\int_0^1 E_{(1-n+\mu),1}^{(\gamma_j)}\left[\frac{-\omega_1(1-n+\mu)}{n-\mu}z^{1-n+\mu},\right.$$

$$\left.\ldots,\frac{-\omega_m(1-n+\mu)}{n-\mu}z^{1-n+\mu}\right]dz =$$

(here $0 < 1-(n-\mu) = 1-n+\mu < 1$)

$$\frac{A(1-n+\mu)}{n-\mu}\int_0^1\left[\sum_{k_1,\ldots,k_m=0}^{\infty}\frac{(\gamma_1)_{k_1}\ldots(\gamma_m)_{k_m}}{\Gamma\left(1+\left(\sum_{j=1}^m k_j\right)(1-n+\mu)\right)}\right.$$

$$\left.\frac{\prod_{j=1}^m\left(\left(\frac{-\omega_j(1-n+\mu)}{n-\mu}\right)^{k_j}z^{(1-n+\mu)\sum_{j=1}^m k_j}\right)}{k_1!\ldots k_m!}\right]dz$$

$$=\left(\frac{A(1-n+\mu)}{n-\mu}\right)$$

$$\sum_{k_1,\ldots,k_m=0}^{\infty}\frac{(\gamma_1)_{k_1}\ldots(\gamma_m)_{k_m}}{\Gamma\left(1+\left(\sum_{j=1}^m k_j\right)(1-n+\mu)\right)}\frac{\prod_{j=1}^m\left(\frac{-\omega_j(1-n+\mu)}{n-\mu}\right)^{k_j}}{k_1!\ldots k_m!}\int_0^1 z^{(1-n+\mu)\sum_{j=1}^m k_j}dz \tag{1.38}$$

$$=\left(\frac{A(1-n+\mu)}{n-\mu}\right)\sum_{k_1,\ldots,k_m=0}^{\infty}\frac{(\gamma_1)_{k_1}\ldots(\gamma_m)_{k_m}}{\Gamma\left(2+\left(\sum_{j=1}^m k_j\right)(1-n+\mu)\right)}\frac{\prod_{j=1}^m\left(\frac{-\omega_j(1-n+\mu)}{n-\mu}\right)^{k_j}}{k_1!\ldots k_m!}$$

$$=\frac{A(1-n+\mu)}{n-\mu}E_{(1-n+\mu),2}^{(\gamma_j)}\left(\frac{-\omega_1(1-n+\mu)}{n-\mu},\ldots,\frac{-\omega_m(1-n+\mu)}{n-\mu}\right)\geq 1, \tag{1.39}$$

for suitable $\omega_j < 0$, for $j = 1, \ldots, m$.

We also need

Definition 1.8 Let $f \in C^n([-1,1])$, $\mathbb{N} \not\ni \mu > 0$, $\lceil\mu\rceil = n \in \mathbb{N}$; $\omega < 0$. That is $0 < 1-n+\mu < 1$. The parametrized Caputo-Fabrizio non-singular kernel fractional derivatives, left and right of order μ, respectively, are given as follows (see also [12]):

$$ {}_{\omega}^{CF}D_{-1+}^{\mu} f(x) := \frac{1}{n-\mu} \int_{-1}^{x} \exp\left(-\frac{(1-n+\mu)\,\omega}{n-\mu}(x-t)\right) f^{(n)}(t)\,dt, \quad (1.40) $$

$$ {}_{\omega}^{CF}D_{1-}^{\mu} f(x) := \frac{(-1)^n}{n-\mu} \int_{x}^{1} \exp\left(-\frac{(1-n+\mu)\,\omega}{n-\mu}(t-x)\right) f^{(n)}(t)\,dt, \quad (1.41) $$

$\forall\, x \in [-1, 1]$.

Equations (1.40), (1.41) are special cases of (1.7), (1.8).

We make

Remark 1.9 We want to evaluate

$$ \infty > \int_{0}^{1} \exp\left(-\frac{(1-n+\mu)\,\omega}{n-\mu} z\right) dz $$

(call $\delta := -\frac{(1-n+\mu)\omega}{n-\mu}$)

$$ = \int_{0}^{1} e^{\delta z} dz = \frac{1}{\delta} e^{\delta z}|_0^1 = \frac{1}{\delta}\left(e^{\delta} - 1\right) = \frac{e^{\delta}}{\delta} - \frac{1}{\delta} = \frac{\sum_{k=0}^{\infty} \frac{\delta^k}{k!}}{\delta} - \frac{1}{\delta} \quad (1.42) $$

$$ = \sum_{k=0}^{\infty} \frac{\delta^{k-1}}{k!} = \sum_{k=0}^{\infty} \left(\frac{\left(\frac{1-n+\mu}{n-\mu}\right)^{k-1} (-\omega)^{k-1}}{k!} \right) \geq 1, $$

for suitable $\omega < 0$.

So, again condition (1.11) is fulfilled.

1.3 Main Results

We give

Theorem 1.10 *Let h, k, p be integers, $0 \leq h \leq k \leq p \in \mathbb{N}$ and let f be a real function, $f^{(p)}$ is continuous in $[-1, 1]$ with modulus of continuity $\omega_1\left(f^{(p)}, \delta\right)$, $\delta > 0$. Let $\alpha_j(x)$, $j = h, h+1, ..., k$ be real functions, defined and bounded on $[-1, 1]$ and assume for $x \in [0, 1]$ that $\alpha_h(x)$ is either $\geq$ some number $\alpha > 0$ or $\leq$ some number $\beta < 0$. Let the real numbers $\alpha_0 = 0 < \alpha_1 < 1 < \alpha_2 < 2 < ... < \alpha_p < p$. Here we* *adopt the abstract fractional calculus terminology and assumptions from above. So* ${}^{k_j}D^{\alpha_j}_{*-1}f$ *stands for the abstract left Caputo type fractional derivative of order* α_j *anchored at* -1. *We consider the linear abstract left fractional differential operator*

$$L := \sum_{j=h}^{k} \alpha_j (x) \left[{}^{k_j}D^{\alpha_j}_{*-1} \right] \tag{1.43}$$

and suppose, throughout $[0, 1]$,

$$L(f) \geq 0.$$

Then, for any $n \in \mathbb{N}$, *there exists a real polynomial* $Q_n(x)$ *of degree* $\leq n$ *such that*

$$L(Q_n) \geq 0 \, \textit{throughout} \; [0, 1], \tag{1.44}$$

and

$$\max_{-1 \leq x \leq 1} |f(x) - Q_n(x)| \leq C n^{k-p} \omega_1 \left(f^{(p)}, \frac{1}{n} \right), \tag{1.45}$$

where C *is independent of* n *or* f.

Proof Let $n \in \mathbb{N}$. By Theorem 1.3 given a real function g, with $g^{(p)}$ continuous in $[-1, 1]$, there exists a real polynomial $q_n(x)$ of degree $\leq n$ such that

$$\max_{-1 \leq x \leq 1} \left| g^{(j)}(x) - q_n^{(j)}(x) \right| \leq R_p n^{j-p} \omega_1 \left(g^{(p)}, \frac{1}{n} \right), \tag{1.46}$$

$j = 0, 1, ..., p$, where R_p is independent of n or g.

We notice that ($x \in [-1, 1]$)

$$\left| \left({}^{k_j}D^{\alpha_j}_{*-1} g \right)(x) - \left({}^{k_j}D^{\alpha_j}_{*-1} q_n \right)(x) \right| =$$

$$\left| \int_{-1}^{x} k_j (x-t) g^{(j)}(t) \, dt - \int_{-1}^{x} k_j (x-t) q_n^{(j)}(t) \, dt \right| =$$

$$\left| \int_{-1}^{x} k_j (x-t) \left(g^{(j)}(t) - q_n^{(j)}(t) \right) dt \right| \leq$$

$$\int_{-1}^{x} k_j (x-t) \left| g^{(j)}(t) - q_n^{(j)}(t) \right| dt \overset{(1.46)}{\leq} \tag{1.47}$$

$$\left(\int_{-1}^{x} k_j (x-t)\, dt\right) R_p n^{j-p} \omega_1 \left(g^{(p)}, \frac{1}{n}\right) =$$

$$\left(\int_{0}^{x+1} k_j (z)\, dz\right) R_p n^{j-p} \omega_1 \left(g^{(p)}, \frac{1}{n}\right) \leq \left(\int_{0}^{2} k_j (z)\, dz\right) R_p n^{j-p} \omega_1 \left(g^{(p)}, \frac{1}{n}\right).$$

We have proved that

$$\left|\left({}^{k_j} D_{*-1}^{\alpha_j} g\right)(x) - \left({}^{k_j} D_{*-1}^{\alpha_j} q_n\right)(x)\right| \leq \tag{1.48}$$

$$\left(\int_{0}^{2} k_j (z)\, dz\right) R_p n^{j-p} \omega_1 \left(g^{(p)}, \frac{1}{n}\right), \ \forall\, x \in [-1, 1].$$

That is

$$\max_{-1 \leq x \leq 1} \left|\left({}^{k_j} D_{*-1}^{\alpha_j} g\right)(x) - \left({}^{k_j} D_{*-1}^{\alpha_j} q_n\right)(x)\right| \leq$$

$$\left(\int_{0}^{2} k_j (z)\, dz\right) R_p n^{j-p} \omega_1 \left(g^{(p)}, \frac{1}{n}\right), \ j = 1, ..., p.$$

So we have

$$\max_{-1 \leq x \leq 1} \left|\left({}^{k_j} D_{*-1}^{\alpha_j} g\right)(x) - \left({}^{k_j} D_{*-1}^{\alpha_j} q_n\right)(x)\right| \leq \lambda_j R_p n^{j-p} \omega_1 \left(g^{(p)}, \frac{1}{n}\right), \tag{1.49}$$

where

$$\lambda_j := \int_{0}^{2} k_j (z)\, dz, \ \ j = 1, ..., p. \tag{1.50}$$

Inequality (1.49) is valid when $j = 0$ by (1.46), and we can set $\lambda_0 = 1$.

Put

$$s_j \equiv \sup_{-1 \leq x \leq 1} \left|\alpha_h^{-1}(x)\, \alpha_j (x)\right|, \ j = h, ..., k, \tag{1.51}$$

and

$$\eta_n := R_p \omega_1 \left(f^{(p)}, \frac{1}{n}\right) \left(\sum_{j=h}^{k} s_j \lambda_j n^{j-p}\right). \tag{1.52}$$

I. Suppose, throughout [0, 1], $\alpha_h (x) \geq \alpha > 0$. Let $Q_n (x)$, $x \in [-1, 1]$, be a real polynomial of degree $\leq n$ so that

$$\max_{-1\le x\le 1}\left|{}^{k_j}D_{*-1}^{\alpha_j}\left(f(x)+\eta_n(h!)^{-1}x^h\right)-\left({}^{k_j}D_{*-1}^{\alpha_j}Q_n\right)(x)\right|\overset{(1.49)}{\le}$$

$$\lambda_j R_p n^{j-p}\omega_1\left(f^{(p)},\frac{1}{n}\right), \tag{1.53}$$

$j=0,1,...,p$.

In particular ($j=0$) holds

$$\max_{-1\le x\le 1}\left|\left(f(x)+\eta_n(h!)^{-1}x^h\right)-Q_n(x)\right|\overset{(1.53)}{\le}R_p n^{-p}\omega_1\left(f^{(p)},\frac{1}{n}\right), \tag{1.54}$$

and

$$\max_{-1\le x\le 1}|f(x)-Q_n(x)|\le\eta_n(h!)^{-1}+R_p n^{-p}\omega_1\left(f^{(p)},\frac{1}{n}\right)=$$

$$(h!)^{-1}R_p\omega_1\left(f^{(p)},\frac{1}{n}\right)\left(\sum_{j=h}^{k}s_j\lambda_j n^{j-p}\right)+R_p n^{-p}\omega_1\left(f^{(p)},\frac{1}{n}\right)\le \tag{1.55}$$

$$R_p\omega_1\left(f^{(p)},\frac{1}{n}\right)n^{k-p}\left(1+(h!)^{-1}\sum_{j=h}^{k}s_j\lambda_j\right).$$

That is

$$\max_{-1\le x\le 1}|f(x)-Q_n(x)|\le R_p\left(1+(h!)^{-1}\sum_{j=h}^{k}s_j\lambda_j\right)n^{k-p}\omega_1\left(f^{(p)},\frac{1}{n}\right), \tag{1.56}$$

proving (1.45).

Here

$$L=\sum_{j=h}^{k}\alpha_j(x)\left[{}^{k_j}D_{*-1}^{\alpha_j}\right],$$

and suppose, throughout $[0,1]$, $Lf\ge 0$. So over $0\le x\le 1$, using (1.52) and (1.53), we have

$$\alpha_h^{-1}(x)L(Q_n(x))=\alpha_h^{-1}(x)L(f(x))+\frac{\eta_n}{h!}\,{}^{k_h}D_{*-1}^{\alpha_h}x^h+$$

$$\sum_{j=h}^{k}\alpha_h^{-1}(x)\alpha_j(x)\left[{}^{k_j}D_{*-1}^{\alpha_j}Q_n(x)-{}^{k_j}D_{*-1}^{\alpha_j}f(x)-\frac{\eta_n}{h!}\,{}^{k_j}D_{*-1}^{\alpha_j}x^h\right]\ge \tag{1.57}$$

$$\frac{\eta_n}{h!}\,{}^{k_h}D_{*-1}^{\alpha_h}x^h - \left(\sum_{j=h}^{k} s_j\lambda_j n^{j-p}\right) R_p\omega_1\left(f^{(p)}, \frac{1}{n}\right) = \frac{\eta_n}{h!}\,{}^{k_h}D_{*-1}^{\alpha_h}x^h - \eta_n =: \varphi$$

(if $h = 0$, then $\alpha_h = 0$, and $\varphi = 0$).

If $h \neq 0$, then

$$\varphi = \eta_n\left(\frac{{}^{k_h}D_{*-1}^{\alpha_h}x^h}{h!} - 1\right) = \eta_n\left(\int_{-1}^{x} k_h(x-t)\,dt - 1\right) =$$

$$\eta_n\left(\int_{0}^{x+1} k_h(z)\,dz - 1\right) \geq \eta_n\left(\int_{0}^{1} k_h(z)\,dz - 1\right) \geq 0 \tag{1.58}$$

by the assumption (1.11): $\int_0^1 k_h(z)\,dz \geq 1$, when $h \neq 0$.

Hence in both cases we get

$$L(Q_n(x)) \geq 0,\ x \in [0, 1]. \tag{1.59}$$

II. Suppose, throughout $[0, 1]$, $\alpha_h(x) \leq \beta < 0$. In this case let $Q_n(x)$, $x \in [-1, 1]$, be a real polynomial of degree $\leq n$ such that

$$\max_{-1\leq x\leq 1}\left|{}^{k_j}D_{*-1}^{\alpha_j}\left(f(x) - \eta_n(h!)^{-1}x^h\right) - \left({}^{k_j}D_{*-1}^{\alpha_j}Q_n\right)(x)\right| \overset{(1.49)}{\leq}$$

$$\lambda_j R_p n^{j-p}\omega_1\left(f^{(p)}, \frac{1}{n}\right), \tag{1.60}$$

$j = 0, 1, \ldots, p$.

In particular ($j = 0$) holds

$$\max_{-1\leq x\leq 1}\left|\left(f(x) - \eta_n(h!)^{-1}x^h\right) - Q_n(x)\right| \overset{(1.60)}{\leq} R_p n^{-p}\omega_1\left(f^{(p)}, \frac{1}{n}\right), \tag{1.61}$$

and

$$\max_{-1\leq x\leq 1}|f(x) - Q_n(x)| \leq \eta_n(h!)^{-1} + R_p n^{-p}\omega_1\left(f^{(p)}, \frac{1}{n}\right) \overset{\text{(as before)}}{\leq} \tag{1.62}$$

$$R_p\omega_1\left(f^{(p)}, \frac{1}{n}\right) n^{k-p}\left(1 + (h!)^{-1}\sum_{j=h}^{k} s_j\lambda_j\right).$$

That is (1.45) is again true.

Again suppose, throughout [0, 1], $Lf \geq 0$. Also if $0 \leq x \leq 1$, then

$$\alpha_h^{-1}(x) L(Q_n(x)) = \alpha_h^{-1}(x) L(f(x)) - \frac{\eta_n}{h!} {}^{k_h}D_{*-1}^{\alpha_h} x^h +$$

$$\sum_{j=h}^{k} \alpha_h^{-1}(x) \alpha_j(x) \left[{}^{k_j}D_{*-1}^{\alpha_j} Q_n(x) - {}^{k_j}D_{*-1}^{\alpha_j} f(x) + \frac{\eta_n}{h!} {}^{k_j}D_{*-1}^{\alpha_j} x^h \right] \overset{(1.60)}{\leq}$$

$$-\frac{\eta_n}{h!} {}^{k_h}D_{*-1}^{\alpha_h} x^h + \left(\sum_{j=h}^{k} s_j \lambda_j n^{j-p} \right) R_p \omega_1 \left(f^{(p)}, \frac{1}{n} \right) = -\frac{\eta_n}{h!} {}^{k_h}D_{*-1}^{\alpha_h} x^h + \eta_n =: \psi$$

(if $h = 0$, then $\alpha_h = 0$, and $\psi = 0$).

If $h \neq 0$, then

$$\psi = \eta_n \left[1 - \frac{{}^{k_h}D_{*-1}^{\alpha_h} x^h}{h!} \right] = \eta_n \left[1 - \int_{-1}^{x} k_h(x-t)\, dt \right] = \tag{1.63}$$

$$\eta_n \left[1 - \int_0^{x+1} k_h(z)\, dz \right] \leq \eta_n \left[1 - \int_0^{1} k_h(z)\, dz \right] \leq 0.$$

Hence again in both cases

$$L(Q_n(x)) \geq 0, \ \forall\, x \in [0, 1]. \tag{1.64}$$

■

We also present

Theorem 1.11 *Let h, k, p be integers, $0 \leq h \leq k \leq p \in \mathbb{N}$, where h is even, and let f be a real function, $f^{(p)}$ is continuous in $[-1, 1]$ with modulus of continuity $\omega_1\left(f^{(p)}, \delta\right)$, $\delta > 0$. Let $\alpha_j(x)$, $j = h, h+1, ..., k$ be real functions, defined and bounded on $[-1, 1]$ and assume for $x \in [-1, 0]$ that $\alpha_h(x)$ is either $\geq$ some number $\alpha > 0$ or $\leq$ some number $\beta < 0$. Let the real numbers $\alpha_0 = 0 < \alpha_1 < 1 < \alpha_2 < 2 < ... < \alpha_p < p$. Here we adopt the abstract fractional calculus terminology and assumptions from above. So ${}^{k_j}D_{1-}^{\alpha_j} f$ stands for the abstract right Caputo type fractional derivative of order α_j anchored at 1. We consider the linear abstract right fractional differential operator*

$$L := \sum_{j=h}^{k} \alpha_j(x) \left[{}^{k_j}D_{1-}^{\alpha_j} \right] \tag{1.65}$$

and suppose, throughout $[-1, 0]$,

$$L\left(f\right)\geq 0.$$

Then, for any $n\in\mathbb{N}$, there exists a real polynomial $Q_n\left(x\right)$ of degree $\leq n$ such that

$$L\left(Q_n\right)\geq 0\,\textit{throughout}\,\left[-1,0\right], \tag{1.66}$$

and

$$\max_{-1\leq x\leq 1}\left|f\left(x\right)-Q_n\left(x\right)\right|\leq Cn^{k-p}\omega_1\left(f^{(p)},\frac{1}{n}\right), \tag{1.67}$$

where C is independent of n or f.

Proof Let $x\in[-1,1]$, we observe that

$$\left|\left({}^{k_j}D_{1-}^{\alpha_j}g\right)(x)-\left({}^{k_j}D_{1-}^{\alpha_j}q_n\right)(x)\right|=$$

$$\left|\int_x^1 k_j\left(t-x\right)g^{(j)}\left(t\right)dt-\int_x^1 k_j\left(t-x\right)q_n^{(j)}\left(t\right)dt\right|=$$

$$\left|\int_x^1 k_j\left(t-x\right)\left(g^{(j)}\left(t\right)-q_n^{(j)}\left(t\right)\right)dt\right|\leq$$

$$\int_x^1 k_j\left(t-x\right)\left|g^{(j)}\left(t\right)-q_n^{(j)}\left(t\right)\right|dt\overset{(1.6)}{\leq} \tag{1.68}$$

$$\left(\int_x^1 k_j\left(t-x\right)dt\right)R_pn^{j-p}\omega_1\left(g^{(p)},\frac{1}{n}\right)=$$

$$\left(\int_0^{1-x} k_j\left(z\right)dz\right)R_pn^{j-p}\omega_1\left(g^{(p)},\frac{1}{n}\right)\leq\left(\int_0^{2} k_j\left(z\right)dz\right)R_pn^{j-p}\omega_1\left(g^{(p)},\frac{1}{n}\right).$$

That is we have derived

$$\max_{-1\leq x\leq 1}\left|\left({}^{k_j}D_{1-}^{\alpha_j}g\right)(x)-\left({}^{k_j}D_{1-}^{\alpha_j}q_n\right)(x)\right|\leq$$

$$\left(\int_0^{2} k_j\left(z\right)dz\right)R_pn^{j-p}\omega_1\left(g^{(p)},\frac{1}{n}\right),\ j=1,...,p. \tag{1.69}$$

We call

$$\lambda_j := \int_0^2 k_j(z)\,dz, \quad j = 1, ..., p. \tag{1.70}$$

Therefore we can write

$$\max_{-1\leq x\leq 1}\left|\left({}^{k_j}D_{1-}^{\alpha_j}g\right)(x) - \left({}^{k_j}D_{1-}^{\alpha_j}q_n\right)(x)\right| \leq \lambda_j R_p n^{j-p}\omega_1\left(g^{(p)}, \frac{1}{n}\right), \tag{1.71}$$

for $j = 1, ..., p$.

Inequality (1.71) is valid when $j = 0$ by (1.6), so we can set $\lambda_0 = 1$.

Put

$$s_j \equiv \sup_{-1\leq x\leq 1}\left|\alpha_h^{-1}(x)\,\alpha_j(x)\right|, \; j = h, ..., k, \tag{1.72}$$

and

$$\eta_n := R_p\omega_1\left(f^{(p)}, \frac{1}{n}\right)\left(\sum_{j=h}^{k} s_j\lambda_j n^{j-p}\right). \tag{1.73}$$

I. Suppose, throughout $[-1, 0]$, $\alpha_h(x) \geq \alpha > 0$. Let $Q_n(x)$, $x \in [-1, 1]$, be a real polynomial of degree $\leq n$ so that

$$\max_{-1\leq x\leq 1}\left|{}^{k_j}D_{1-}^{\alpha_j}\left(f(x) + \eta_n(h!)^{-1}x^h\right) - \left({}^{k_j}D_{1-}^{\alpha_j}Q_n\right)(x)\right| \overset{(1.71)}{\leq}$$

$$\lambda_j R_p n^{j-p}\omega_1\left(f^{(p)}, \frac{1}{n}\right), \tag{1.74}$$

$j = 0, 1, ..., p$.

In particular $(j = 0)$ holds

$$\max_{-1\leq x\leq 1}\left|\left(f(x) + \eta_n(h!)^{-1}x^h\right) - Q_n(x)\right| \overset{(1.74)}{\leq} R_p n^{-p}\omega_1\left(f^{(p)}, \frac{1}{n}\right), \tag{1.75}$$

and, as earlier,

$$\max_{-1\leq x\leq 1}\left|f(x) - Q_n(x)\right| \leq R_p\left(1 + (h!)^{-1}\sum_{j=h}^{k} s_j\lambda_j\right)n^{k-p}\omega_1\left(f^{(p)}, \frac{1}{n}\right), \tag{1.76}$$

proving (1.67).

Here

$$L = \sum_{j=h}^{k}\alpha_j(x)\left[{}^{k_j}D_{1-}^{\alpha_j}\right],$$

and suppose, throughout $[-1, 0]$, $Lf \geq 0$. So over $-1 \leq x \leq 0$, we get

$$\alpha_h^{-1}(x) L(Q_n(x)) = \alpha_h^{-1}(x) L(f(x)) + \frac{\eta_n}{h!} {}^{k_h}D_{1-}^{\alpha_h} x^h +$$

$$\sum_{j=h}^{k} \alpha_h^{-1}(x) \alpha_j(x) \left[{}^{k_j}D_{1-}^{\alpha_j} Q_n(x) - {}^{k_j}D_{1-}^{\alpha_j} f(x) - \frac{\eta_n}{h!} {}^{k_j}D_{1-}^{\alpha_j} x^h \right] \overset{(1.74)}{\geq} \tag{1.77}$$

$$\frac{\eta_n}{h!} {}^{k_h}D_{1-}^{\alpha_h} x^h - \left(\sum_{j=h}^{k} s_j \lambda_j n^{j-p} \right) R_p \omega_1 \left(f^{(p)}, \frac{1}{n} \right) = \frac{\eta_n}{h!} {}^{k_h}D_{1-}^{\alpha_h} x^h - \eta_n =: \xi$$

(if $h = 0$, then $\alpha_h = 0$, and $\xi = 0$).

If $h \neq 0$, then

$$\xi = \eta_n \left(\frac{{}^{k_h}D_{1-}^{\alpha_h} x^h}{h!} - 1 \right) = \eta_n \left((-1)^h \int_x^1 k_h(t - x)\, dt - 1 \right) =$$

(h is even)

$$\eta_n \left(\int_x^1 k_h(t - x)\, dt - 1 \right) = \eta_n \left(\int_0^{1-x} k_h(z)\, dz - 1 \right) \geq$$

$$\eta_n \left(\int_0^1 k_h(z)\, dz - 1 \right) \geq 0, \tag{1.78}$$

by the assumption (1.11): $\int_0^1 k_h(z)\, dz \geq 1$, when $h \neq 0$.

Hence in both cases we get

$$L(Q_n(x)) \geq 0, \; x \in [-1, 0]. \tag{1.79}$$

II. Suppose, throughout $[-1, 0]$, $\alpha_h(x) \leq \beta < 0$. Let $Q_n(x)$, $x \in [-1, 1]$, be a real polynomial of degree $\leq n$ so that

$$\max_{-1 \leq x \leq 1} \left| {}^{k_j}D_{1-}^{\alpha_j} \left(f(x) - \eta_n (h!)^{-1} x^h \right) - \left({}^{k_j}D_{1-}^{\alpha_j} Q_n \right)(x) \right| \overset{(1.71)}{\leq}$$

$$\lambda_j R_p n^{j-p} \omega_1 \left(f^{(p)}, \frac{1}{n} \right), \tag{1.80}$$

$j = 0, 1, ..., p$.

In particular ($j = 0$) holds

$$\max_{-1\le x\le 1}\left|\left(f(x)-\eta_n (h!)^{-1}x^h\right)-Q_n(x)\right| \overset{(1.80)}{\le} R_p n^{-p}\omega_1\left(f^{(p)},\frac{1}{n}\right), \tag{1.81}$$

and as earlier,

$$\max_{-1\le x\le 1}|f(x)-Q_n(x)|\le R_p\left(1+(h!)^{-1}\sum_{j=h}^{k}s_j\lambda_j\right)n^{k-p}\omega_1\left(f^{(p)},\frac{1}{n}\right), \tag{1.82}$$

proving (1.67).

Again suppose, throughout $[-1, 0]$, $Lf \ge 0$. Also if $-1 \le x \le 0$, then

$$\alpha_h^{-1}(x)L(Q_n(x)) = \alpha_h^{-1}(x)L(f(x)) - \frac{\eta_n}{h!}\,{}^{k_h}D_{1-}^{\alpha_h}x^h +$$

$$\sum_{j=h}^{k}\alpha_h^{-1}(x)\alpha_j(x)\left[{}^{k_j}D_{1-}^{\alpha_j}Q_n(x) - {}^{k_j}D_{1-}^{\alpha_j}f(x) + \frac{\eta_n}{h!}\,{}^{k_j}D_{1-}^{\alpha_j}x^h\right] \overset{(1.80)}{\le} \tag{1.83}$$

$$-\frac{\eta_n}{h!}\,{}^{k_h}D_{1-}^{\alpha_h}x^h + \left(\sum_{j=h}^{k}s_j\lambda_j n^{j-p}\right)R_p\omega_1\left(f^{(p)},\frac{1}{n}\right) = -\frac{\eta_n}{h!}\,{}^{k_h}D_{1-}^{\alpha_h}x^h + \eta_n$$

$$= \eta_n\left(1-\frac{{}^{k_h}D_{1-}^{\alpha_h}x^h}{h!}\right) =: \rho$$

(if $h = 0$, then $\alpha_h = 0$, and $\rho = 0$).

If $h \ne 0$, then

$$\rho = \eta_n\left(1-\int_x^1 k_h(t-x)\,dt\right) = \eta_n\left(1-\int_0^{1-x}k_h(z)\,dz\right) \le \tag{1.84}$$

$$\eta_n\left(1-\int_0^1 k_h(z)\,dz\right) \le 0.$$

Hence in both cases we get again

$$L(Q_n(x)) \ge 0,\ \forall\, x \in [-1, 0]\,. \tag{1.85}$$

Conclusion 1.12 *Clearly Theorem 1.10 generalizes Theorem 1.2, and Theorem 1.11 generalizes Theorem 2.2, p. 12 of [2]. Furthermore there, the approximating polynomial Q_n depends on f, η_n, h; which η_n depends on n, $R_p, n, k, s_j, \lambda_j$; which λ_j depends on k_j. I.e. polynomial Q_n among others depends on the type of fractional calculus we use.*

Consequently, Theorem 1.10 is valid for the following left fractional linear differential operators:

(1)

$$L_1 := \sum_{j=h}^{k} \alpha_j (x) \left[{}^{C}D^{\gamma}_{\rho,\alpha_j,\omega,-1+} \right], \tag{1.86}$$

where $\rho > 0$, $\gamma < 0$, and $\omega > 0$ large enough (from Prabhakar fractional calculus, see (1.16));

(2)

$$L_2 := \sum_{j=h}^{k} \alpha_j (x) \left[D^{\alpha_j}_{-1*} \right], \tag{1.87}$$

(see (1.35)) where $\gamma_j > 0$, $j = 1, ..., m$; $\lambda = 1$; and small enough $\omega_j < 0$, $j = 1, ..., m$ (from generalized non-singular fractional calculus);

and

(3)

$$L_3 := \sum_{j=h}^{k} \alpha_j (x) \left[{}^{CF}_{\omega} D^{\alpha_j}_{-1+} \right], \tag{1.88}$$

with $\omega < 0$, sufficiently small (from parametrized Caputo-Fabrizio non-singular kernel fractional calculus).

Similarly, Theorem 1.11 is valid for the following right fractional linear differential operators:

*(1)**

$$L_1^* := \sum_{j=h}^{k} \alpha_j (x) \left[{}^{C}D^{\gamma}_{\rho,\alpha_j,\omega,1-} \right], \tag{1.89}$$

where $\rho > 0$, $\gamma < 0$, and $\omega > 0$ large enough (from Prabhakar fractional calculus, see (1.17));

*(2)**

$$L_2^* := \sum_{j=h}^{k} \alpha_j (x) \left[D^{\alpha_j}_{1-} \right], \tag{1.90}$$

(see (1.36)) where $\gamma_j > 0$, $j = 1, ..., m$; $\lambda = 1$; and small enough $\omega_j < 0$, $j = 1, ..., m$ (from generalized non-singular fractional calculus);

and

*(3)**

$$L_3^* := \sum_{j=h}^{k} \alpha_j(x) \left[{}_{\omega}^{CF} D_{1-}^{\alpha_j}\right], \tag{1.91}$$

with $\omega < 0$, sufficiently small (from parametrized Caputo-Fabrizio non-singular kernel fractional calculus).

Our developed abstract fractional monotone approximation theory with its applications involves weaker conditions than the one with ordinary derivatives ([6]), and can cover many diverse general cases in a multitude of complex settings and environments.

References

1. Anastassiou, G.A.: Bivariate Monotone Approximation. Proc. Amer. Math. Soc. **112**(4), 959–964 (1991)
2. Anastassiou, G.A.: Frontiers in Approximation Theory. World Scientific Publishing Co Pte Ltd., New Jersey, Singapore (2015)
3. Anastassiou, G.A.: Foundations of Generalized Prabhakar-Hilfer fractional Calculus with Applications. Cubo (2021, accepted)
4. Anastassiou, G.A.: Multiparameter fractional differentiation with non singular kernel. Issues Anal (2021, accepted)
5. Anastassiou, G.A.: Abstract fractional monotone approximation with applications (2021, submitted)
6. Anastassiou, G.A., Shisha, O.: Monotone approximation with linear differential operators. J. Approx. Theor. **44**, 391–393 (1985)
7. Atangana, A., Baleanu, D.: New fractional derivatives with nonlocal and non-singular kernel: theory and application to heat transfer model. Therm. Sci. **20**(2), 763–769 (2016)
8. Diethelm, K.: The Analysis of Fractional Differential Equations, Lecture Notes in Mathematics, Vol. 2004, 1st edn. Springer, New York, Heidelberg (2010)
9. Giusti, A., et al.: A practical guide to Prabhakar fractional calculus. Fractional Calc. Appl. Anal. **23**(1), 9–54 (2020)
10. Gorenflo, R., Kilbas, A., Mainardi, F., Rogosin, S.: Mittag-Leffler Functions. Related Topics and Applications. Springer, Heidelberg, New York (2014)
11. Hewitt, E., Stromberg, K.: Real and Abstract Analysis. Springer, New York, Heidelberg, Berlin (1965)
12. Losada, J., Nieto, J.J.: Properties of a new fractional derivative without singular kernel. Progr. Fract. Differ. Appl. **1**(2), 87–92 (2015)
13. Polito, F., Tomovski, Z.: Some properties of Prabhakar-type fractional calculus operators. Fractional Differ. Calc. **6**(1), 73–94 (2016)
14. Saxena, R.K., Kalla, S.L., Saxena, R.: Integr. Transf. Spec. Funct. Multivariate analogue of generalized Mittag-Leffler function **22**(7), 533–548 (2011)
15. Shisha, O.: Monotone approximation. Pacific J. Math. **15**, 667–671 (1965)
16. Srivastava, H.M., Daoust, M.C.: A note on the convergence of Kompe' de Feriet's double hypergeometric series. Math. Nachr. **53**, 151–159 (1972)
17. Teljakovskii, S.A.: Two theorems on the approximation of functions by algebraic polynomials. Mat. Sb. **70**(112), 252–265 (1966) [Russian]; Amer. Math. Soc. Trans. **77**(2), 163–178 (1968)
18. Trigub, R.M.: Approximation of functions by polynomials with integer coefficients. Izv. Akad. Nauk SSSR Ser. Mat. 26, 261–280 (1962) [Russian]

Chapter 2
Advanced Abstract Fractional Monotone Approximation

Here we extend our earlier univariate high order simultaneous fractional monotone approximation theory to abstract univariate high order simultaneous fractional monotone approximation, with applications to Prabhakar fractional calculus and non-singular kernel fractional calculi. We cover both the left and right sides of this constrained approximation. Let $f \in C^r([-1, 1])$, $r \geq 0$ and let L^* be a linear abstract left or right fractional differential operator such that $L^*(f) \geq 0$ over $[0, 1]$ or $[-1, 0]$, respectively. We can find a sequence of polynomials Q_n of degree $\leq n$ such that $L^*(Q_n) \geq 0$ over $[0, 1]$ or $[-1, 0]$, furthermore f is approximated left or right fractionally and simultaneously by Q_n on $[-1, 1]$. The degree of these restricted approximations is given quantitatively by inequalities using a higher order modulus of smoothness for $f^{(r)}$. It follows [6].

2.1 Introduction

The topic of monotone approximation started in 1965, [17] by O. Shisha, and became a major part of approximation theory. The original problem was: given a positive integer k, approximate with rates a given function f whose kth derivative is ≥ 0 by polynomials $(Q_n)_{n\in\mathbb{N}}$ having the same property.

In 1985, [7], the author and O. Shisha, continued this study by replacing the kth derivative with a linear differential operator of order k involving ordinary derivatives, again the approximation was with rates.

Later in 1991, [1], the author extended this kind of study in two dimensions, etc.

We use also the notation $I = [-1, 1]$.

We would like to mention

G. A. Anastassiou, *Abstract Fractional Monotone Approximation, Theory and Applications*, Studies in Systems, Decision and Control 411,
https://doi.org/10.1007/978-3-030-95943-2_2

Theorem 2.1 *(Gonska and Hinnemann [11]). Let $r \geq 0$ and $s \geq 1$. Then there exists a sequence $Q_n = Q_n^{(r,s)}$ of linear polynomial operators mapping $C^r(I)$ into P_n (space of polynomials of degree $\leq n$), such that for all $f \in C^r(I)$, all $|x| \leq 1$ and all $n \geq \max(4(r+1), r+s)$ we have*

$$\left|f^{(k)}(x) - (Q_n f)^{(k)}(x)\right| \leq M_{r,s}(\Delta_n(x))^{r-k}\,\omega_s\left(f^{(r)}, \Delta_n(x)\right),\ 0 \leq k \leq r, \tag{2.1}$$

where $\Delta_n(x) = \frac{\sqrt{1-x^2}}{n} + \frac{1}{n^2}$, and $M_{r,s}$ is a constant independent of f, x, and n. Above ω_n is the usual modulus of smoothness of order s with respect to the supremum norm.

Theorem 2.1 implies the useful

Corollary 2.2 *([2]) Let $r \geq 0$ and $s \geq 1$. Then there exists a sequence $Q_n = Q_n^{(r,s)}$ of linear polynomial operators mapping $C^{(r)}(I)$ into P_n, such that for all $f \in C^r(I)$ and all $n \geq \max(4(r+1), r+s)$ we have*

$$\left\|f^{(k)} - (Q_n f)^{(k)}\right\|_\infty \leq \frac{C_{r,s}}{n^{r-k}}\omega_s\left(f^{(r)}, \frac{1}{n}\right),\ k = 0, 1, \ldots, r, \tag{2.2}$$

where $C_{r,s}$ is a constant independent of f and n.

In [2] we proved the motivational

Theorem 2.3 *Let h, v, r be integers, $0 \leq h \leq v \leq r$ and let $f \in C^r(I)$, with $f^{(r)}$ having modulus of smoothness $\omega_s\left(f^{(r)}, \delta\right)$ there, $s \geq 1$. Let $\alpha_j(x)$, $j = h, h+1, \ldots, v$ be real functions, defined and bounded on I and suppose α_h is either $\geq \alpha > 0$ or $\leq \beta < 0$ throughout I. Take the operator*

$$L = \sum_{j=h}^{v} \alpha_j(x)\left[\frac{d^j}{dx^j}\right] \tag{2.3}$$

and assume, throughout I,

$$L(f) \geq 0. \tag{2.4}$$

Then, for every integer $n \geq \max(4(r+1), r+s)$, there exists a real polynomial $Q_n(x)$ of degree $\leq n$ such that

$$L(Q_n) \geq 0 \text{ throughout } I, \tag{2.5}$$

and

$$\left\|f^{(k)} - Q_n^{(k)}\right\|_\infty \leq \frac{C}{n^{r-v}}\omega_s\left(f^{(r)}, \frac{1}{n}\right),\ 0 \leq k \leq h. \tag{2.6}$$

Moreover, we set

$$\left\| f^{(k)} - Q_n^{(k)} \right\|_\infty \leq \frac{C}{n^{r-k}} \omega_s \left(f^{(r)}, \frac{1}{n} \right), \; h+1 \leq k \leq r, \tag{2.7}$$

where C is a constant independent of f and n.

In [3], Chap. 3, we extended Theorem 2.3 to the fractional level. Indeed there L is replaced by L^*, a linear left Caputo fractional differential operator. Then the monotonicity property is only true on the critical interval $[0, 1]$. Simultaneous and fractional convergence remained true on all of I.

We need

Definition 2.4 ([9], p. 50) Let $\alpha > 0$ and $\lceil \alpha \rceil = m$, ($\lceil \cdot \rceil$ ceiling of the number). Consider $f \in C^m([-1, 1])$. We define the left Caputo fractional derivative of f of order α as follows:

$$\left(D_{*-1}^{\alpha} f \right)(x) = \frac{1}{\Gamma(m-\alpha)} \int_{-1}^{x} (x-t)^{m-\alpha-1} f^{(m)}(t) \, dt, \tag{2.8}$$

for any $x \in [-1, 1]$, where Γ is the gamma function.

We set

$$\begin{gathered} D_{*-1}^{0} f(x) = f(x), \\ D_{*-1}^{m} f(x) = f^{(m)}(x), \; \forall \, x \in [-1, 1]. \end{gathered} \tag{2.9}$$

So in [3], chap. 3, we proved the following result:

Theorem 2.5 *Let h, v, r be integers, $1 \leq h \leq v \leq r$ and let $f \in C^r([-1, 1])$, with $f^{(r)}$ having modulus of smoothness $\omega_s\left(f^{(r)}, \delta\right)$ there, $s \geq 1$. Let $\alpha_j(x)$, $j = h, h+1, \ldots, v$ be real functions, defined and bounded on $[-1, 1]$ and suppose $\alpha_h(x)$ is either $\geq \alpha > 0$ or $\leq \beta < 0$ on $[0, 1]$. Let the real numbers $\alpha_0 = 0 < \alpha_1 \leq 1 < \alpha_2 \leq 2 < \ldots < \alpha_r \leq r$. Here $D_{*-1}^{\alpha_j} f$ stands for the left Caputo fractional derivative of f of order α_j anchored at -1. Consider the linear left fractional differential operator*

$$L^* := \sum_{j=h}^{v} \alpha_j(x) \left[D_{*-1}^{\alpha_j} \right] \tag{2.10}$$

and suppose, throughout $[0, 1]$,

$$L^*(f) \geq 0. \tag{2.11}$$

Then, for any $n \in \mathbb{N}$ such that $n \geq \max(4(r+1), r+s)$, there exists a real polynomial $Q_n(x)$ of degree $\leq n$ such that

$$L^*(Q_n) \geq 0 \text{ throughout } [0, 1], \tag{2.12}$$

and

$$\sup_{-1\leq x\leq 1}\left|\left(D_{*-1}^{\alpha_j}f\right)(x)-\left(D_{*-1}^{\alpha_j}Q_n\right)(x)\right|\leq \frac{2^{j-\alpha_j}}{\Gamma\left(j-\alpha_j+1\right)}\frac{C_{r,s}}{n^{r-j}}\omega_s\left(f^{(r)},\frac{1}{n}\right), \tag{2.13}$$

$j=h+1,...,r$; $C_{r,s}$ *is a constant independent of* f *and* n.

Set

$$l_j:\equiv \sup_{x\in[-1,1]}\left|\alpha_h^{-1}(x)\,\alpha_j(x)\right|,\ h\leq j\leq v. \tag{2.14}$$

When $j=1,...,h$ *we derive*

$$\sup_{-1\leq x\leq 1}\left|\left(D_{*-1}^{\alpha_j}f\right)(x)-\left(D_{*-1}^{\alpha_j}Q_n\right)(x)\right|\leq \frac{C_{r,s}}{n^{r-v}}\omega_s\left(f^{(r)},\frac{1}{n}\right)\cdot$$

$$\left[\left(\sum_{\tau=h}^{v}l_\tau\frac{2^{\tau-\alpha_\tau}}{\Gamma\left(\tau-\alpha_\tau+1\right)}\right)\left(\sum_{\lambda=0}^{h-j}\frac{2^{h-\alpha_j-\lambda}}{\lambda!\Gamma\left(h-\alpha_j-\lambda+1\right)}\right)+\frac{2^{j-\alpha_j}}{\Gamma\left(j-\alpha_j+1\right)}\right]. \tag{2.15}$$

Finally it holds

$$\sup_{-1\leq x\leq 1}\left|f(x)-Q_n(x)\right|\leq \frac{C_{r,s}}{n^{r-v}}\omega_s\left(f^{(r)},\frac{1}{n}\right)\left[\frac{1}{h!}\sum_{\tau=h}^{v}l_\tau\frac{2^{\tau-\alpha_\tau}}{\Gamma\left(\tau-\alpha_\tau+1\right)}+1\right]. \tag{2.16}$$

In this chapter we establish abstract fractional calculus left and right high order monotone approximation theory of Caputo type, and then we apply our results to Prabhakar fractional Calculus, generalized non-singular fractional calculus, and parametrized Caputo-Fabrizio non-singular fractional calculus.

Next, we build the related necessary fractional calculi background.

2.2 Fractional Calculi

Here we need to be very specific in preparation for our main results.

2.2.1 *Abstract Fractional Calculus*

Let $h,v\in\mathbb{Z}_+$, $r\in\mathbb{N}:0\leq h\leq v\leq r$. Let also $\mathbb{N}\not\ni\alpha_j>0$, $j=1,...,r$, such that $\alpha_0=0<\alpha_1<1<\alpha_2<2<\alpha_3<3<...<...<\alpha_r<r$. That is $\lceil\alpha_j\rceil=j$, $j=1,...,r$; $\lceil\alpha_0\rceil=0$.

Consider the integrable functions $k_j := K_{\alpha_j} : [0, 2] \rightarrow \mathbb{R}_+$, $j = 0, 1, ..., r$. Here $g \in C^r([-1, 1])$.

We consider the following abstract left side Caputo type fractional derivatives:

$$\left({}^{k_j}D_{*-1}^{\alpha_j} g\right)(x) := \int_{-1}^{x} k_j(x-t)\, g^{(j)}(t)\, dt, \tag{2.17}$$

$j = 1, .., r$; $\forall\, x \in [-1, 1]$.

Similarly, we define the corresponding right side generalized Caputo type fractional derivatives:

$$\left({}^{k_j}D_{1-}^{\alpha_j} g\right)(x) := (-1)^j \int_{x}^{1} k_j(t-x)\, g^{(j)}(t)\, dt, \tag{2.18}$$

$j = 1, .., r$; $\forall\, x \in [-1, 1]$.

We set

$$\left({}^{k_j}D_{*-1}^{j} g\right)(x) := g^{(j)}(x); \ \left({}^{k_j}D_{1-}^{j} g\right)(x) := (-1)^j g^{(j)}(x), \tag{2.19}$$

for $j = 1, ..., r$, and also we set

$$\left({}^{k_0}D_{*-1}^{0} g\right)(x) := \left({}^{k_0}D_{1-}^{0} g\right)(x) := g(x), \tag{2.20}$$

$\forall\, x \in [-1, 1]$.

We will assume that

$$\int_{0}^{1} k_h(z)\, dz \geq 1, \text{ when } h \neq 0. \tag{2.21}$$

In the usual Caputo fractional derivatives case it is

$$k_j(z) = \frac{z^{j-\alpha_j-1}}{\Gamma\left(j-\alpha_j\right)}, \ j = 1, ..., r; \ \forall\, z \in [0, 2], \tag{2.22}$$

and (2.21) is fulfilled, by the fact that $\Gamma(h - \alpha_h + 1) \leq 1$, see [3], p. 6.

2.2.2 *About Prabhakar Fractional Calculus*

Here we follow [4, 15].

We consider the Prabhakar function (also known as the three parameter Mittag-Leffler function), (see [12], p. 97; [10])

$$E_{\alpha,\beta}^{\gamma}(z)=\sum_{k=0}^{\infty}\frac{(\gamma)_k}{k!\Gamma(\alpha k+\beta)}z^k, \tag{2.23}$$

where Γ is the gamma function; $\alpha,\beta>0$, $\gamma\in\mathbb{R}$, $z\in\mathbb{R}$, and $(\gamma)_k=\gamma(\gamma+1)\ldots(\gamma+k-1)$. It is $E_{\alpha,\beta}^{0}(z)=\frac{1}{\Gamma(\beta)}$.

Let $a,b\in\mathbb{R}$, $a<b$ and $x\in[a,b]$; $f\in C([a,b])$. The left and right Prabhakar fractional integrals are defined ([4, 15]) as follows:

$$\left(e_{\rho,\mu,\omega,a+}^{\gamma}f\right)(x)=\int_a^x(x-t)^{\mu-1}E_{\rho,\mu}^{\gamma}\left[\omega(x-t)^{\rho}\right]f(t)\,dt, \tag{2.24}$$

and

$$\left(e_{\rho,\mu,\omega,b-}^{\gamma}f\right)(x)=\int_x^b(t-x)^{\mu-1}E_{\rho,\mu}^{\gamma}\left[\omega(t-x)^{\rho}\right]f(t)\,dt, \tag{2.25}$$

where $\rho,\mu>0$; $\gamma,\omega\in\mathbb{R}$.

Functions (2.24) and (2.25) are continuous, see [4].

Next, let $f\in C^N([a,b])$, where $N=\lceil\mu\rceil$, ($\lceil\cdot\rceil$ is the ceiling of the number), $0<\mu\notin\mathbb{N}$. We define the Prabhakar-Caputo left and right fractional derivatives of order μ ([4, 15]) as follows ($x\in[a,b]$):

$$\left({}^{C}D_{\rho,\mu,\omega,a+}^{\gamma}f\right)(x)=\int_a^x(x-t)^{N-\mu-1}E_{\rho,N-\mu}^{-\gamma}\left[\omega(x-t)^{\rho}\right]f^{(N)}(t)\,dt, \tag{2.26}$$

with ${}^{C}D_{\rho,0,\omega,a+}^{\gamma}f:=f$; ${}^{C}D_{\rho,N,\omega,a+}^{\gamma}f:=f^{(N)}$, $N\in\mathbb{N}$,
and

$$\left({}^{C}D_{\rho,\mu,\omega,b-}^{\gamma}f\right)(x)=(-1)^N\int_x^b(t-x)^{N-\mu-1}E_{\rho,N-\mu}^{-\gamma}\left[\omega(t-x)^{\rho}\right]f^{(N)}(t)\,dt, \tag{2.27}$$

with ${}^{C}D_{\rho,0,\omega,b-}^{\gamma}f:=f$; ${}^{C}D_{\rho,N,\omega,b-}^{\gamma}f:=(-1)^Nf^{(N)}$, $N\in\mathbb{N}$.

One can rewrite (2.26) and (2.27) as follows:

$$\left({}^{C}D_{\rho,\mu,\omega,a+}^{\gamma}f\right)(x)=\left(e_{\rho,N-\mu,\omega,a+}^{-\gamma}f^{[N]}\right)(x), \tag{2.28}$$

and

$$\left({}^{C}D_{\rho,\mu,\omega,b-}^{\gamma}f\right)(x)=(-1)^N\left(e_{\rho,N-\mu,\omega,b-}^{-\gamma}f^{[N]}\right)(x), \tag{2.29}$$

$\forall x \in [a, b]$.

Clearly, the functions (2.28) and (2.29) are continuous.

2.2.3 From Generalized Non-singular Fractional Calculus ([5])

We need

Definition 2.6 Here we use the multivariate analogue of generalized Mittag-Leffler function, see [16], defined for $\lambda, \gamma_j, \rho_j, z_j \in \mathbb{C}$, $Re\left(\rho_j\right) > 0\,(j = 1, ..., m)$ in terms of a multiple series of the form:

$$E_{(\rho_j),\lambda}^{(\gamma_j)}(z_1, ..., z_m) = E_{(\rho_1,...,\rho_m),\lambda}^{(\gamma_1,...,\gamma_m)}(z_1, ..., z_m) =$$

$$\sum_{k_1,...,k_m=0}^{\infty} \frac{(\gamma_1)_{k_1} \cdots (\gamma_m)_{k_m}}{\Gamma\left(\lambda + \sum_{j=1}^{m} k_j \rho_j\right)} \frac{z_1^{k_1} ... z_m^{k_m}}{k_1! ... k_m!}, \tag{2.30}$$

where $\left(\gamma_j\right)_{k_j}$ is the Pochhammer symbol, Γ is the gamma function. By [18], p. 157, (2.30) converges for $Re\left(\rho_j\right) > 0$, $j = 1, ..., m$.

In what follows we will use the particular case of $E_{(\rho,...,\rho),\lambda}^{(\gamma_1,...,\gamma_m)}[\omega_1 t^\rho, ..., \omega_m t^\rho]$, denoted by $E_{(\rho),\lambda}^{(\gamma_j)}[\omega_1 t^\rho, ..., \omega_m t^\rho]$, where $0 < \rho < 1$, $t \geq 0$, $\lambda > 0$, $\gamma_j \in \mathbb{R}$ with $\left(\gamma_j\right)_{k_j} := \gamma_j \left(\gamma_j + 1\right) ... \left(\gamma_j + k_j - 1\right)$, $\omega_j \in \mathbb{R} - \{0\}$, for $j = 1, ..., m$.

Let now $f \in C^{n+1}([a, b])$, $n \in \mathbb{Z}_+$.

We define the Caputo type generalized left fractional derivative with non-singular kernel of order $n + \rho$, as

$$D_{a*}^{n+\rho} f(x) := {}_{(\gamma_j)(\omega_j)}^{CA} D_{a*}^{n+\rho,\lambda} f(x) :=$$

$$\frac{A(\rho)}{1-\rho} \int_a^x E_{(\rho),\lambda}^{(\gamma_j)}\left[\frac{-\omega_1 \rho}{1-\rho}(x-t)^\rho, ..., \frac{-\omega_m \rho}{1-\rho}(x-t)^\rho\right] f^{(n+1)}(t)\, dt, \tag{2.31}$$

$\forall x \in [a, b]$.

Similarly, we define the Caputo type generalized right fractional derivative with non-singular kernel of order $n + \rho$, as

$$D_{b-}^{n+\rho} f(x) := {}_{(\gamma_j)(\omega_j)}^{CA} D_{b-}^{n+\rho,\lambda} f(x) :=$$

$$(-1)^{n+1} \frac{A(\rho)}{1-\rho} \int_x^b E_{(\rho),\lambda}^{(\gamma_j)} \left[\frac{-\omega_1 \rho}{1-\rho} (t-x)^{\rho}, ..., \frac{-\omega_m \rho}{1-\rho} (t-x)^{\rho} \right] f^{(n+1)}(t) \, dt, \tag{2.32}$$

$\forall x \in [a, b]$.

Above $A(\rho)$ is a normalizing constant.

The above derivatives (2.31), (2.32) generalize the Atangana-Baleanu fractional derivatives [8].

We rewrite (2.31) and (2.32), and for $[a, b] = [-1, 1]$.

Let $\mu > 0$ with $\mu \notin \mathbb{N}$ and $\lceil \mu \rceil = n \in \mathbb{N}$. That is $0 < 1 - n + \mu < 1$, and let $f \in C^n([-1, 1])$. Then, we have

$$D_{-1*}^{\mu} f(x) := {}_{(\gamma_j)(\omega_j)}^{CA} D_{-1*}^{\mu,\lambda} f(x) :=$$

$$\frac{A(1-n+\mu)}{n-\mu} \int_{-1}^{x} E_{(1-n+\mu),\lambda}^{(\gamma_j)} \left[\frac{-\omega_1 (1-n+\mu)}{n-\mu} (x-t)^{1-n+\mu}, ..., \right. \tag{2.33}$$

$$\left. \frac{-\omega_m (1-n+\mu)}{n-\mu} (x-t)^{1-n+\mu} \right] f^{(n)}(t) \, dt,$$

and

$$D_{1-}^{\mu} f(x) := {}_{(\gamma_j)(\omega_j)}^{CA} D_{1-}^{\mu,\lambda} f(x) :=$$

$$(-1)^n \frac{A(1-n+\mu)}{n-\mu} \int_{x}^{1} E_{(1-n+\mu),\lambda}^{(\gamma_j)} \left[\frac{-\omega_1 (1-n+\mu)}{n-\mu} (t-x)^{1-n+\mu}, ..., \right. \tag{2.34}$$

$$\left. \frac{-\omega_m (1-n+\mu)}{n-\mu} (t-x)^{1-n+\mu} \right] f^{(n)}(t) \, dt,$$

$\forall x \in [-1, 1]$.

We will set $D_{-1*}^{0} f = f$, $D_{1-}^{0} f = f$, and $D_{-1*}^{m} f = f^{(m)}$, $D_{1-}^{m} f = (-1)^m f^{(m)}$, when $m \in \mathbb{N}$.

We make

Remark 2.7 Fractional Calculi of Sects. 2.2.2 and 2.2.3 are special cases of abstract fractional calculus, see Sect. 2.2.1. In particular the important condition (2.21) is fulfilled.

So, we will verify $\int_0^1 k_h(z) \, dz \geq 1$, $h \neq 0$.

(I) First for Sect. 2.2.2:

We notice that

$$\int_0^1 z^{N-\mu-1} E_{\rho,N-\mu}^{-\gamma}\left(\omega z^{\rho}\right) dz =$$

(here $\rho, N-\mu > 0, \gamma < 0, \omega > 0$)

$$\int_0^1 z^{N-\mu-1} \sum_{k=0}^{\infty} \frac{(-\gamma)_k}{k!\Gamma(\rho k + N - \mu)} \left(\omega z^{\rho}\right)^k dz \overset{\text{(by [13], p. 175)}}{=}$$

$$\sum_{k=0}^{\infty} \frac{(-\gamma)_k \, \omega^k}{k!\Gamma(\rho k + N - \mu)} \int_0^1 z^{N-\mu-1} z^{\rho k} dz = \tag{2.35}$$

$$\sum_{k=0}^{\infty} \frac{(-\gamma)_k \, \omega^k}{k!\Gamma(\rho k + N - \mu)} \int_0^1 z^{(\rho k+N)-\mu-1} dz =$$

$$\sum_{k=0}^{\infty} \frac{(-\gamma)_k \, \omega^k}{k!\Gamma(\rho k + N - \mu + 1)} = E_{\rho,N-\mu+1}^{-\gamma}(\omega) \geq 1,$$

for suitable $\omega > 0$.

(II) Next, for Sect. 2.2.3:

Here $\gamma_j > 0, j = 1, \ldots, m; \lambda = 1; \mathbb{N} \not\ni \mu > 0, \lceil \mu \rceil = n \in \mathbb{N}, \omega_j < 0, j = 1, \ldots, m$.

Without loss of generality we assume that $A(1-n+\mu) > 0$.

We have that

$$\frac{A(1-n+\mu)}{n-\mu} \int_0^1 E_{(1-n+\mu),1}^{(\gamma_j)} \left[\frac{-\omega_1 (1-n+\mu)}{n-\mu} z^{1-n+\mu}, \right.$$

$$\left. \ldots, \frac{-\omega_m (1-n+\mu)}{n-\mu} z^{1-n+\mu} \right] dz =$$

(here $0 < 1-(n-\mu) = 1-n+\mu < 1$)

$$\frac{A(1-n+\mu)}{n-\mu} \int_0^1 \left[\sum_{k_1,\ldots,k_m=0}^{\infty} \frac{(\gamma_1)_{k_1} \cdots (\gamma_m)_{k_m}}{\Gamma\left(1 + \left(\sum_{j=1}^{m} k_j\right)(1-n+\mu)\right)} \right.$$

$$\left.\frac{\prod_{j=1}^{m}\left(\left(\frac{-\omega_j(1-n+\mu)}{n-\mu}\right)^{k_j} z^{(1-n+\mu)\sum_{j=1}^{m}k_j}\right)}{k_1!\ldots k_m!}\right] dz$$

$$= \left(\frac{A(1-n+\mu)}{n-\mu}\right)$$

$$\sum_{k_1,\ldots,k_m=0}^{\infty} \frac{(\gamma_1)_{k_1}\ldots(\gamma_m)_{k_m}}{\Gamma\left(1+\left(\sum_{j=1}^{m}k_j\right)(1-n+\mu)\right)} \frac{\prod_{j=1}^{m}\left(\frac{-\omega_j(1-n+\mu)}{n-\mu}\right)^{k_j}}{k_1!\ldots k_m!} \int_0^1 z^{(1-n+\mu)\sum_{j=1}^{m}k_j} dz$$

$$= \left(\frac{A(1-n+\mu)}{n-\mu}\right) \sum_{k_1,\ldots,k_m=0}^{\infty} \frac{(\gamma_1)_{k_1}\ldots(\gamma_m)_{k_m}}{\Gamma\left(2+\left(\sum_{j=1}^{m}k_j\right)(1-n+\mu)\right)} \frac{\prod_{j=1}^{m}\left(\frac{-\omega_j(1-n+\mu)}{n-\mu}\right)^{k_j}}{k_1!\ldots k_m!} \tag{2.36}$$

$$= \frac{A(1-n+\mu)}{n-\mu} E_{(1-n+\mu),2}^{(\gamma_j)}\left(\frac{-\omega_1(1-n+\mu)}{n-\mu},\ldots,\frac{-\omega_m(1-n+\mu)}{n-\mu}\right) \geq 1, \tag{2.37}$$

for suitable $\omega_j < 0$, for $j = 1, \ldots, m$.

We also need

Definition 2.8 Let $f \in C^n([-1,1])$, $\mathbb{N} \not\ni \mu > 0$, $\lceil \mu \rceil = n \in \mathbb{N}$; $\omega < 0$. That is $0 < 1 - n + \mu < 1$. The parametrized Caputo-Fabrizio non-singular kernel fractional derivatives, left and right of order μ, respectively, are given as follows (see also [14]):

$${}^{CF}_{\omega}D^{\mu}_{-1+}f(x) := \frac{1}{n-\mu}\int_{-1}^{x} \exp\left(-\frac{(1-n+\mu)\,\omega}{n-\mu}(x-t)\right) f^{(n)}(t)\,dt, \tag{2.38}$$

$${}^{CF}_{\omega}D^{\mu}_{1-}f(x) := \frac{(-1)^n}{n-\mu}\int_{x}^{1} \exp\left(-\frac{(1-n+\mu)\,\omega}{n-\mu}(t-x)\right) f^{(n)}(t)\,dt, \tag{2.39}$$

$\forall\, x \in [-1, 1]$.

Equations (2.38), (2.39) are special cases of (2.17), (2.18).

We make

Remark 2.9 We want to calculate

$$\infty > \int_0^1 \exp\left(-\frac{(1-n+\mu)\,\omega}{n-\mu}z\right)dz$$

(set $\delta := -\frac{(1-n+\mu)\omega}{n-\mu}$)

$$= \int_0^1 e^{\delta z}dz = \frac{1}{\delta}e^{\delta z}|_0^1 = \frac{1}{\delta}\left(e^{\delta}-1\right) = \frac{e^{\delta}}{\delta} - \frac{1}{\delta} = \frac{\sum_{k=0}^{\infty}\frac{\delta^k}{k!}}{\delta} - \frac{1}{\delta} \tag{2.40}$$

$$= \sum_{k=0}^{\infty}\frac{\delta^{k-1}}{k!} = \sum_{k=0}^{\infty}\left(\frac{\left(\frac{1-n+\mu}{n-\mu}\right)^{k-1}(-\omega)^{k-1}}{k!}\right) \geq 1,$$

for suitable $\omega < 0$.

So, again condition (2.21) is valid.

2.3 Main Results

We present

Theorem 2.10 *Let h, v, r be integers, $0 \leq h \leq v \leq r$ and let $f \in C^r([-1,1])$, with $f^{(r)}$ having modulus of smoothness $\omega_s\left(f^{(r)}, \delta\right)$ there, $s \geq 1$. Let $\alpha_j(x)$, $j = h, h+1, ..., v$ be real functions, defined and bounded on $[-1,1]$ and suppose $\alpha_h(x)$ is either $\geq \alpha > 0$ or $\leq \beta < 0$ on $[0,1]$. Let the real numbers $\alpha_0 = 0 < \alpha_1 < 1 < \alpha_2 < 2 < ... < \alpha_r < r$. Here ${}^{k_j}D_{*-1}^{\alpha_j}f$ stands for the abstract left Caputo fractional derivative of f of order α_j anchored at -1. We consider the abstract left fractional linear differential operator*

$$L^* := \sum_{j=h}^{v}\alpha_j(x)\left[{}^{k_j}D_{*-1}^{\alpha_j}\right], \tag{2.41}$$

and suppose, throughout $[0,1]$,

$$L^*(f) \geq 0. \tag{2.42}$$

Set

$$\lambda_j := \int_0^2 k_j(z)\,dz,\ j = 1, ..., r, \tag{2.43}$$

and $\lambda_0 = 1$; along with

$$s_j := \sup_{-1 \leq x \leq 1} \left| \alpha_h^{-1}(x) \alpha_j(x) \right|, \ j = h, ..., k. \tag{2.44}$$

Then, for any $n \in \mathbb{N} : n \geq \max(4(r+1), r+s)$, *there exists a real polynomial* $Q_n(x)$ *of degree* $\leq n$ *such that*

$$L^*(Q_n) \geq 0, \ \textit{throughout} \ [0,1], \tag{2.45}$$

and

(i)

$$\max_{-1 \leq x \leq 1} \left| \left({}^{k_j}D_{*-1}^{\alpha_j} f \right)(x) - \left({}^{k_j}D_{*-1}^{\alpha_j} Q_n \right)(x) \right| \leq \lambda_j \frac{C_{r,s}}{n^{r-j}} \omega_s \left(f^{(r)}, \frac{1}{n} \right), \tag{2.46}$$

when $j = h+1, ..., r$*;*

(ii) if $h \neq 0$ *and* $j = 1, ..., h$*, we have*

$$\max_{-1 \leq x \leq 1} \left| \left({}^{k_j}D_{*-1}^{\alpha_j} f \right)(x) - \left({}^{k_j}D_{*-1}^{\alpha_j} Q_n \right)(x) \right| \leq$$

$$\frac{C_{r,s} \omega_s \left(f^{(r)}, \frac{1}{n} \right)}{n^{r-v}} \left[\frac{\left(\sum_{j_*=h}^{v} s_{j_*} \lambda_{j_*} \right)}{h!} \max_{-1 \leq x \leq 1} \left| \left({}^{k_j}D_{*-1}^{\alpha_j} x^h \right) \right| + \lambda_j \right]; \tag{2.47}$$

(iii) furthermore

$$\max_{-1 \leq x \leq 1} |f(x) - Q_n(x)| \leq \frac{C_{r,s} \omega_s \left(f^{(r)}, \frac{1}{n} \right)}{n^{r-v}} \left[\frac{\left(\sum_{j=h}^{v} s_j \lambda_j \right)}{h!} + 1 \right]. \tag{2.48}$$

Proof Let Q_n be as in Corollary 2.2. Let also $\alpha_j > 0$, $j = 1, ..., r$, such that $\alpha_0 = 0 < \alpha_1 < 1 < \alpha_2 < 2 < ... < \alpha_r < r$. That is $\lceil \alpha_j \rceil = j$, $j = 0, 1, ..., r$.

We consider the abstract left Caputo fractional derivatives ${}^{k_j}D_{*-1}^{\alpha_j} f$ and ${}^{k_j}D_{*-1}^{\alpha_j} Q_n$, $j = 1, ..., r$.

We notice that ($x \in [-1, 1]$)

$$\left| \left({}^{k_j}D_{*-1}^{\alpha_j} f \right)(x) - \left({}^{k_j}D_{*-1}^{\alpha_j} Q_n \right)(x) \right| =$$

$$\left| \int_{-1}^{x} k_j(x-t) f^{(j)}(t) \, dt - \int_{-1}^{x} k_j(x-t) Q_n^{(j)}(t) \, dt \right| = \tag{2.49}$$

$$\left|\int_{-1}^{x} k_j(x-t)\left(f^{(j)}(t) - Q_n^{(j)}(t)\right)dt\right| \leq$$

$$\int_{-1}^{x} k_j(x-t)\left|f^{(j)}(t) - Q_n^{(j)}(t)\right|dt \overset{(2.2.)}{\leq}$$

$$\left(\int_{-1}^{x} k_j(x-t)\,dt\right)\frac{C_{r,s}}{n^{r-j}}\omega_s\left(f^{(r)}, \frac{1}{n}\right) =$$

$$\left(\int_{0}^{x+1} k_j(z)\,dz\right)\frac{C_{r,s}}{n^{r-j}}\omega_s\left(f^{(r)}, \frac{1}{n}\right) \leq \left(\int_{0}^{2} k_j(z)\,dz\right)\frac{C_{r,s}}{n^{r-j}}\omega_s\left(f^{(r)}, \frac{1}{n}\right). \tag{2.50}$$

We have proved that

$$\left|\left({}^{k_j}D_{*-1}^{\alpha_j}f\right)(x) - \left({}^{k_j}D_{*-1}^{\alpha_j}Q_n\right)(x)\right| \leq$$

$$\left(\int_{0}^{2} k_j(z)\,dz\right)\frac{C_{r,s}}{n^{r-j}}\omega_s\left(f^{(r)}, \frac{1}{n}\right), \tag{2.51}$$

$\forall\, x \in [-1, 1]$, $j = 1, ..., r$.

Hence it holds

$$\max_{-1\leq x\leq 1}\left|\left({}^{k_j}D_{*-1}^{\alpha_j}f\right)(x) - \left({}^{k_j}D_{*-1}^{\alpha_j}Q_n\right)(x)\right| \leq \tag{2.52}$$

$$\lambda_j\frac{C_{r,s}}{n^{r-j}}\omega_s\left(f^{(r)}, \frac{1}{n}\right),\ j = 1, ..., r,$$

where

$$\lambda_j := \int_{0}^{2} k_j(z)\,dz,\ j = 1, ..., r,\ \text{as in (2.43.)}.$$

Inequality (2.52) is valid when $j = 0$ by (2.2) and (2.9), and we can set $\lambda_0 = 1$.

Set

$$\rho_n := C_{r,s}\omega_s\left(f^{(r)}, \frac{1}{n}\right)\left(\sum_{j=h}^{v} s_j\lambda_j n^{j-r}\right). \tag{2.53}$$

I. Suppose, throughout [0, 1], $\alpha_h(x) \geq \alpha > 0$.

Let $Q_n(x)$, $x \in [-1, 1]$, be a real polynomial of degree $\leq n$ as in Corollary 2.2, so that

$$\max_{-1 \leq x \leq 1} \left| {}^{k_j} D_{*-1}^{\alpha_j} \left(f(x) + \rho_n (h!)^{-1} x^h \right) - \left({}^{k_j} D_{*-1}^{\alpha_j} Q_n \right)(x) \right| \overset{(2.52.)}{\leq} \tag{2.54}$$

$$\lambda_j \frac{C_{r,s}}{n^{r-j}} \omega_s \left(f^{(r)}, \frac{1}{n} \right), \ j = 0, 1, ..., r.$$

When $j = h + 1, ..., r$, we get

$$\max_{-1 \leq x \leq 1} \left| \left({}^{k_j} D_{*-1}^{\alpha_j} f \right)(x) - \left({}^{k_j} D_{*-1}^{\alpha_j} Q_n \right)(x) \right| \leq \lambda_j \frac{C_{r,s}}{n^{r-j}} \omega_s \left(f^{(r)}, \frac{1}{n} \right), \tag{2.55}$$

proving (2.46).

For $j = 1, ..., h \neq 0$, we have that

$$\max_{-1 \leq x \leq 1} \left| \left({}^{k_j} D_{*-1}^{\alpha_j} f \right)(x) - \left({}^{k_j} D_{*-1}^{\alpha_j} Q_n \right)(x) \right| \overset{(2.54.)}{\leq}$$

$$\frac{\rho_n}{h!} \max_{-1 \leq x \leq 1} \left| \left({}^{k_j} D_{*-1}^{\alpha_j} x^h \right) \right| + \lambda_j \frac{C_{r,s}}{n^{r-j}} \omega_s \left(f^{(r)}, \frac{1}{n} \right) =$$

$$\frac{C_{r,s}}{h!} \omega_s \left(f^{(r)}, \frac{1}{n} \right) \left(\sum_{j_*=h}^{v} s_{j_*} \lambda_{j_*} n^{j_*-r} \right) \max_{-1 \leq x \leq 1} \left| \left({}^{k_j} D_{*-1}^{\alpha_j} x^h \right) \right| + \lambda_j \frac{C_{r,s}}{n^{r-j}} \omega_s \left(f^{(r)}, \frac{1}{n} \right)$$

$$= C_{r,s} \omega_s \left(f^{(r)}, \frac{1}{n} \right) \left[\frac{1}{h!} \left(\sum_{j_*=h}^{v} s_{j_*} \lambda_{j_*} n^{j_*-r} \right) \max_{-1 \leq x \leq 1} \left| \left({}^{k_j} D_{*-1}^{\alpha_j} x^h \right) \right| + \frac{\lambda_j}{n^{r-j}} \right] \leq \tag{2.56}$$

$$\frac{C_{r,s} \omega_s \left(f^{(r)}, \frac{1}{n} \right)}{n^{r-v}} \left[\frac{\left(\sum_{j_*=h}^{v} s_{j_*} \lambda_{j_*} \right)}{h!} \max_{-1 \leq x \leq 1} \left| \left({}^{k_j} D_{*-1}^{\alpha_j} x^h \right) \right| + \lambda_j \right],$$

proving (2.47).

When $j = 0$ from (2.54) we get

$$\max_{-1 \leq x \leq 1} \left| f(x) + \rho_n (h!)^{-1} x^h - Q_n(x) \right| \leq \frac{C_{r,s}}{n^r} \omega_s \left(f^{(r)}, \frac{1}{n} \right). \tag{2.57}$$

That is

$$\max_{-1 \leq x \leq 1} |f(x) - Q_n(x)| \leq \frac{\rho_n}{h!} + \frac{C_{r,s}}{n^r} \omega_s \left(f^{(r)}, \frac{1}{n} \right) = \tag{2.58}$$

$$\frac{C_{r,s}\omega_s\left(f^{(r)},\frac{1}{n}\right)}{h!}\left(\sum_{j=h}^{v}s_j\lambda_j n^{j-r}\right)+\frac{C_{r,s}}{n^r}\omega_s\left(f^{(r)},\frac{1}{n}\right)\le$$

$$C_{r,s}\omega_s\left(f^{(r)},\frac{1}{n}\right)\left[\frac{1}{h!n^{r-v}}\left(\sum_{j=h}^{v}s_j\lambda_j\right)+\frac{1}{n^r}\right]\le \tag{2.59}$$

$$\frac{C_{r,s}\omega_s\left(f^{(r)},\frac{1}{n}\right)}{n^{r-v}}\left[\frac{\left(\sum_{j=h}^{v}s_j\lambda_j\right)}{h!}+1\right],$$

proving (2.48).

Also if $0 \le x \le 1$, then

$$\alpha_h^{-1}(x)\,L^*\left(Q_n(x)\right)=\alpha_h^{-1}(x)\,L^*\left(f(x)\right)+\frac{\rho_n}{h!}\,{}^{k_h}D_{*-1}^{\alpha_h}x^h+ \tag{2.60}$$

$$\sum_{j=h}^{v}\alpha_h^{-1}(x)\,\alpha_j(x)\left[{}^{k_j}D_{*-1}^{\alpha_j}Q_n(x)-{}^{k_j}D_{*-1}^{\alpha_j}f(x)-\frac{\rho_n}{h!}\,{}^{k_j}D_{*-1}^{\alpha_j}x^h\right]\overset{(2.54.)}{\ge}$$

(by $L^*f \ge 0$)

$$\frac{\rho_n}{h!}\,{}^{k_h}D_{*-1}^{\alpha_h}x^h-C_{r,s}\omega_s\left(f^{(r)},\frac{1}{n}\right)\left(\sum_{j=h}^{v}s_j\lambda_j n^{j-r}\right)=\frac{\rho_n}{h!}\,{}^{k_h}D_{*-1}^{\alpha_h}x^h-\rho_n=:\varphi$$

(if $h=0$, then $\alpha_h=0$, and $\varphi=0$).

If $h \ne 0$, then

$$\varphi=\rho_n\left(\frac{{}^{k_h}D_{*-1}^{\alpha_h}x^h}{h!}-1\right)=\rho_n\left(\int_{-1}^{x}k_h(x-t)\,dt-1\right)= \tag{2.61}$$

$$\rho_n\left(\int_0^{x+1}k_h(z)\,dz-1\right)\ge\rho_n\left(\int_0^{1}k_h(z)\,dz-1\right)\ge 0,$$

by the assumption (2.21): $\int_0^1 k_h(z)\,dz \ge 1$, when $h \ne 0$.

Hence in both cases we get

$$L^*\left(Q_n(x)\right)\ge 0,\ x\in[0,1]. \tag{2.62}$$

II. Suppose on $[0,1]$ that $\alpha_h(x)\le\beta<0$.

Let $Q_n(x)$, $x\in[-1,1]$, be a real polynomial of degree $\le n$ as in Corollary 2.2, such that

$$\max_{-1 \leq x \leq 1} \left| {}^{k_j}D_{*-1}^{\alpha_j} \left(f(x) - \rho_n (h!)^{-1} x^h \right) - \left({}^{k_j}D_{*-1}^{\alpha_j} Q_n \right)(x) \right| \overset{(2.52.)}{\leq} \tag{2.63}$$

$$\lambda_j \frac{C_{r,s}}{n^{r-j}} \omega_s \left(f^{(r)}, \frac{1}{n} \right), \ j = 0, 1, ..., r.$$

Similarly, we obtain again the inequalities of convergence, see (2.46), (2.47) and (2.48).

Also if $0 \leq x \leq 1$, then

$$\alpha_h^{-1}(x) L^*(Q_n(x)) = \alpha_h^{-1}(x) L^*(f(x)) - \frac{\rho_n}{h!} {}^{k_h}D_{*-1}^{\alpha_h} x^h +$$

$$\sum_{j=h}^{v} \alpha_h^{-1}(x) \alpha_j(x) \left[{}^{k_j}D_{*-1}^{\alpha_j} Q_n(x) - {}^{k_j}D_{*-1}^{\alpha_j} f(x) + \frac{\rho_n}{h!} {}^{k_j}D_{*-1}^{\alpha_j} x^h \right] \overset{(2.63.)}{\leq}$$

(by $L^* f \geq 0$)

$$-\frac{\rho_n}{h!} {}^{k_h}D_{*-1}^{\alpha_h} x^h + C_{r,s} \omega_s \left(f^{(r)}, \frac{1}{n} \right) \left(\sum_{j=h}^{v} s_j \lambda_j n^{j-r} \right) = -\frac{\rho_n}{h!} {}^{k_h}D_{*-1}^{\alpha_h} x^h + \rho_n =: \psi \tag{2.64}$$

(if $h = 0$, then $\alpha_h = 0$, and $\psi = 0$).

If $h \neq 0$, then

$$\psi = \rho_n \left[1 - \frac{{}^{k_h}D_{*-1}^{\alpha_h} x^h}{h!} \right] = \rho_n \left[1 - \int_{-1}^{x} k_h(x - t) \, dt \right] =$$

$$\rho_n \left[1 - \int_{0}^{x+1} k_h(z) \, dz \right] \leq \rho_n \left[1 - \int_{0}^{1} k_h(z) \, dz \right] \leq 0.$$

Hence again in both cases it holds

$$L^*(Q_n(x)) \geq 0, \ \forall x \in [0, 1]. \tag{2.65}$$

We also give

Theorem 2.11 *Let h, v, r be integers, h is even, $0 \leq h \leq v \leq r$ and let $f \in C^r([-1, 1])$, with $f^{(r)}$ having modulus of smoothness $\omega_s\left(f^{(r)}, \delta\right)$ there, $s \geq 1$. Let $\alpha_j(x)$, $j = h, h+1, ..., v$ be real functions, defined and bounded on $[-1, 1]$ and suppose $\alpha_h(x)$ is either $\geq \alpha > 0$ or $\leq \beta < 0$ on $[-1, 0]$. Let the real numbers $\alpha_0 = 0 < \alpha_1 < 1 < \alpha_2 < 2 < ... < \alpha_r < r$. Here ${}^{k_j}D_{1-}^{\alpha_j} f$ stands for the abstract right Caputo fractional derivative of f of order α_j anchored at 1. We consider the abstract right fractional linear differential operator*

$$L^* := \sum_{j=h}^{v} \alpha_j(x)\left[{}^{k_j}D_{1-}^{\alpha_j}\right], \tag{2.66}$$

and suppose, throughout $[-1, 0]$,

$$L^*(f) \geq 0. \tag{2.67}$$

Set

$$\lambda_j := \int_0^2 k_j(z)\,dz,\ j = 1, ..., r, \tag{2.68}$$

and $\lambda_0 = 1$*; along with*

$$s_j := \sup_{-1 \leq x \leq 1} \left|\alpha_h^{-1}(x)\,\alpha_j(x)\right|,\ j = h, ..., k. \tag{2.69}$$

Then, for any $n \in \mathbb{N} : n \geq \max(4(r+1), r+s)$*, there exists a real polynomial* $Q_n(x)$ *of degree* $\leq n$ *such that*

$$L^*(Q_n) \geq 0,\ \textit{throughout}\ [-1, 0], \tag{2.70}$$

and

(i)

$$\max_{-1 \leq x \leq 1} \left|\left({}^{k_j}D_{1-}^{\alpha_j}f\right)(x) - \left({}^{k_j}D_{1-}^{\alpha_j}Q_n\right)(x)\right| \leq \lambda_j \frac{C_{r,s}}{n^{r-j}}\omega_s\left(f^{(r)}, \frac{1}{n}\right), \tag{2.71}$$

when $j = h+1, ..., r$*;*

(ii) if $h \neq 0$ *and* $j = 1, ..., h$*, we have*

$$\max_{-1 \leq x \leq 1} \left|\left({}^{k_j}D_{1-}^{\alpha_j}f\right)(x) - \left({}^{k_j}D_{1-}^{\alpha_j}Q_n\right)(x)\right| \leq$$

$$\frac{C_{r,s}\omega_s\left(f^{(r)}, \frac{1}{n}\right)}{n^{r-v}}\left[\frac{\left(\sum_{j_*=h}^{v} s_{j_*}\lambda_{j_*}\right)}{h!}\max_{-1 \leq x \leq 1}\left|\left({}^{k_j}D_{1-}^{\alpha_j}x^h\right)\right| + \lambda_j\right]; \tag{2.72}$$

(iii) furthermore

$$\max_{-1 \leq x \leq 1} |f(x) - Q_n(x)| \leq \frac{C_{r,s}\omega_s\left(f^{(r)}, \frac{1}{n}\right)}{n^{r-v}}\left[\frac{\left(\sum_{j=h}^{v} s_j\lambda_j\right)}{h!} + 1\right]. \tag{2.73}$$

Proof Let Q_n be as in Corollary 2.2. Let also $\alpha_j > 0$, $j = 1, ..., r$, such that $\alpha_0 = 0 < \alpha_1 < 1 < \alpha_2 < 2 < ... < \alpha_r < r$. That is $\lceil \alpha_j \rceil = j$, $j = 0, 1, ..., r$.

We consider the abstract right Caputo fractional derivatives ${}^{k_j}D_{1-}^{\alpha_j} f$ and ${}^{k_j}D_{1-}^{\alpha_j} Q_n$, $j = 1, ..., r$.

We notice that ($x \in [-1, 1]$)

$$\left| \left({}^{k_j}D_{1-}^{\alpha_j} f \right)(x) - \left({}^{k_j}D_{1-}^{\alpha_j} Q_n \right)(x) \right| =$$

$$\left| (-1)^j \int_x^1 k_j (t - x) f^{(j)} (t) \, dt - (-1)^j \int_x^1 k_j (t - x) Q_n^{(j)} (t) \, dt \right| = \tag{2.74}$$

$$\left| \int_x^1 k_j (t - x) \left(f^{(j)} (t) - Q_n^{(j)} (t) \right) dt \right| \leq$$

$$\int_x^1 k_j (t - x) \left| f^{(j)} (t) - Q_n^{(j)} (t) \right| dt \overset{(2.2.)}{\leq}$$

$$\left(\int_x^1 k_j (t - x) \right) \frac{C_{r,s}}{n^{r-j}} \omega_s \left(f^{(r)}, \frac{1}{n} \right) =$$

$$\left(\int_0^{1-x} k_j (z) \, dz \right) \frac{C_{r,s}}{n^{r-j}} \omega_s \left(f^{(r)}, \frac{1}{n} \right) \leq \left(\int_0^2 k_j (z) \, dz \right) \frac{C_{r,s}}{n^{r-j}} \omega_s \left(f^{(r)}, \frac{1}{n} \right). \tag{2.75}$$

We have proved that

$$\left| \left({}^{k_j}D_{1-}^{\alpha_j} f \right)(x) - \left({}^{k_j}D_{1-}^{\alpha_j} Q_n \right)(x) \right| \leq$$

$$\left(\int_0^2 k_j (z) \, dz \right) \frac{C_{r,s}}{n^{r-j}} \omega_s \left(f^{(r)}, \frac{1}{n} \right), \tag{2.76}$$

$\forall \, x \in [-1, 1]$, $j = 1, ..., r$.

Hence it holds

$$\max_{-1 \leq x \leq 1} \left| \left({}^{k_j}D_{1-}^{\alpha_j} f \right)(x) - \left({}^{k_j}D_{1-}^{\alpha_j} Q_n \right)(x) \right| \leq \tag{2.77}$$

$$\lambda_j \frac{C_{r,s}}{n^{r-j}} \omega_s \left(f^{(r)}, \frac{1}{n} \right), \quad j = 1, ..., r,$$

where

$$\lambda_j := \int_0^2 k_j(z)\,dz, \ j = 1, ..., r, \text{ as in (2.68).}$$

Inequality (2.77) is valid when $j = 0$ by (2.2) and (2.9), and we can set $\lambda_0 = 1$.

Set

$$\rho_n := C_{r,s}\omega_s\left(f^{(r)}, \frac{1}{n}\right)\left(\sum_{j=h}^{v} s_j\lambda_j n^{j-r}\right). \tag{2.78}$$

I. Suppose, throughout $[-1, 0]$, $\alpha_h(x) \geq \alpha > 0$.

Let $Q_n(x)$, $x \in [-1, 1]$, be a real polynomial of degree $\leq n$ as in Corollary 2.2, so that

$$\max_{-1\leq x\leq 1}\left|{}^{k_j}D_{1-}^{\alpha_j}\left(f(x) + \rho_n (h!)^{-1} x^h\right) - \left({}^{k_j}D_{1-}^{\alpha_j}Q_n\right)(x)\right| \overset{(2.77.)}{\leq} \tag{2.79}$$

$$\lambda_j \frac{C_{r,s}}{n^{r-j}}\omega_s\left(f^{(r)}, \frac{1}{n}\right), \ j = 0, 1, ..., r.$$

When $j = h+1, ..., r$, we get

$$\max_{-1\leq x\leq 1}\left|\left({}^{k_j}D_{1-}^{\alpha_j}f\right)(x) - \left({}^{k_j}D_{1-}^{\alpha_j}Q_n\right)(x)\right| \leq \lambda_j \frac{C_{r,s}}{n^{r-j}}\omega_s\left(f^{(r)}, \frac{1}{n}\right), \tag{2.80}$$

proving (2.71).

For $j = 1, ..., h \neq 0$, we have that

$$\max_{-1\leq x\leq 1}\left|\left({}^{k_j}D_{1-}^{\alpha_j}f\right)(x) - \left({}^{k_j}D_{1-}^{\alpha_j}Q_n\right)(x)\right| \overset{(2.79.)}{\leq}$$

$$\frac{\rho_n}{h!}\max_{-1\leq x\leq 1}\left|\left({}^{k_j}D_{1-}^{\alpha_j}x^h\right)\right| + \lambda_j \frac{C_{r,s}}{n^{r-j}}\omega_s\left(f^{(r)}, \frac{1}{n}\right) =$$

$$\frac{C_{r,s}}{h!}\omega_s\left(f^{(r)}, \frac{1}{n}\right)\left(\sum_{j_*=h}^{v} s_{j_*}\lambda_{j_*}n^{j_*-r}\right)\max_{-1\leq x\leq 1}\left|\left({}^{k_j}D_{1-}^{\alpha_j}x^h\right)\right| + \lambda_j \frac{C_{r,s}}{n^{r-j}}\omega_s\left(f^{(r)}, \frac{1}{n}\right)$$

$$= C_{r,s}\omega_s\left(f^{(r)}, \frac{1}{n}\right)\left[\frac{1}{h!}\left(\sum_{j_*=h}^{v} s_{j_*}\lambda_{j_*}n^{j_*-r}\right)\max_{-1\leq x\leq 1}\left|\left({}^{k_j}D_{1-}^{\alpha_j}x^h\right)\right| + \frac{\lambda_j}{n^{r-j}}\right] \leq \tag{2.81}$$

$$\frac{C_{r,s}\omega_s\left(f^{(r)}, \frac{1}{n}\right)}{n^{r-v}}\left[\frac{\left(\sum_{j_*=h}^{v} s_{j_*}\lambda_{j_*}\right)}{h!}\max_{-1\leq x\leq 1}\left|\left({}^{k_j}D_{1-}^{\alpha_j}x^h\right)\right| + \lambda_j\right],$$

proving (2.72).

When $j = 0$ from (2.79) we get

$$\max_{-1 \leq x \leq 1} \left| f(x) + \rho_n (h!)^{-1} x^h - Q_n(x) \right| \leq \frac{C_{r,s}}{n^r} \omega_s \left(f^{(r)}, \frac{1}{n} \right). \tag{2.82}$$

That is

$$\max_{-1 \leq x \leq 1} |f(x) - Q_n(x)| \leq \frac{\rho_n}{h!} + \frac{C_{r,s}}{n^r} \omega_s \left(f^{(r)}, \frac{1}{n} \right) =$$

$$\frac{C_{r,s} \omega_s \left(f^{(r)}, \frac{1}{n} \right)}{h!} \left(\sum_{j=h}^{v} s_j \lambda_j n^{j-r} \right) + \frac{C_{r,s}}{n^r} \omega_s \left(f^{(r)}, \frac{1}{n} \right) \leq \tag{2.83}$$

$$C_{r,s} \omega_s \left(f^{(r)}, \frac{1}{n} \right) \left[\frac{1}{h! n^{r-v}} \left(\sum_{j=h}^{v} s_j \lambda_j \right) + \frac{1}{n^r} \right] \leq$$

$$\frac{C_{r,s} \omega_s \left(f^{(r)}, \frac{1}{n} \right)}{n^{r-v}} \left[\frac{\left(\sum_{j=h}^{v} s_j \lambda_j \right)}{h!} + 1 \right], \tag{2.84}$$

proving (2.73).

Also if $-1 \leq x \leq 0$, then

$$\alpha_h^{-1}(x) L^* (Q_n(x)) = \alpha_h^{-1}(x) L^* (f(x)) + \frac{\rho_n}{h!} {}^{k_h}D_{1-}^{\alpha_h} x^h + \tag{2.85}$$

$$\sum_{j=h}^{v} \alpha_h^{-1}(x) \alpha_j(x) \left[{}^{k_j}D_{1-}^{\alpha_j} Q_n(x) - {}^{k_j}D_{1-}^{\alpha_j} f(x) - \frac{\rho_n}{h!} {}^{k_j}D_{1-}^{\alpha_j} x^h \right] \overset{(2.79.)}{\geq}$$

(by $L^* f \geq 0$)

$$\frac{\rho_n}{h!} {}^{k_h}D_{1-}^{\alpha_h} x^h - C_{r,s} \omega_s \left(f^{(r)}, \frac{1}{n} \right) \left(\sum_{j=h}^{v} s_j \lambda_j n^{j-r} \right) = \frac{\rho_n}{h!} {}^{k_h}D_{1-}^{\alpha_h} x^h - \rho_n =: \varphi$$

(if $h = 0$, then $\alpha_h = 0$, and $\varphi = 0$).

If $h \neq 0$, then

$$\varphi = \rho_n \left(\frac{{}^{k_h}D_{1-}^{\alpha_h} x^h}{h!} - 1 \right) = \rho_n \left((-1)^h \int_x^1 k_h (t - x) \, dt - 1 \right) \overset{(h \text{ is even})}{=} \tag{2.86}$$

$$\rho_n \left(\int_x^1 k_h (t - x)\, dt - 1 \right) = \rho_n \left(\int_0^{1-x} k_h (z)\, dz - 1 \right) \geq$$

$$\rho_n \left(\int_0^1 k_h (z)\, dz - 1 \right) \geq 0,$$

by the assumption (2.21): $\int_0^1 k_h (z)\, dz \geq 1$, when $h \neq 0$.

Hence in both cases we get

$$L^* (Q_n (x)) \geq 0, \ x \in [-1, 0]. \tag{2.87}$$

II. Suppose on $[-1, 0]$ that $\alpha_h (x) \leq \beta < 0$.

Let $Q_n (x)$, $x \in [-1, 1]$, be a real polynomial of degree $\leq n$ as in Corollary 2.2, so that

$$\max_{-1 \leq x \leq 1} \left| {}^{k_j} D_{1-}^{\alpha_j} \left(f (x) - \rho_n (h!)^{-1} x^h \right) - \left({}^{k_j} D_{1-}^{\alpha_j} Q_n \right) (x) \right| \overset{(2.77.)}{\leq} \tag{2.88}$$

$$\lambda_j \frac{C_{r,s}}{n^{r-j}} \omega_s \left(f^{(r)}, \frac{1}{n} \right), \ j = 0, 1, ..., r.$$

Similarly, we obtain again the inequalities of convergence, see (2.71), (2.72) and (2.73).

Also if $-1 \leq x \leq 0$, then

$$\alpha_h^{-1} (x) L^* (Q_n (x)) = \alpha_h^{-1} (x) L^* (f (x)) - \frac{\rho_n}{h!} {}^{k_h} D_{1-}^{\alpha_h} x^h +$$

$$\sum_{j=h}^{v} \alpha_h^{-1} (x) \alpha_j (x) \left[{}^{k_j} D_{1-}^{\alpha_j} Q_n (x) - {}^{k_j} D_{1-}^{\alpha_j} f (x) + \frac{\rho_n}{h!} {}^{k_j} D_{1-}^{\alpha_j} x^h \right] \overset{(2.88.)}{\leq}$$

(by $L^* f \geq 0$)

$$- \frac{\rho_n}{h!} {}^{k_h} D_{1-}^{\alpha_h} x^h + C_{r,s} \omega_s \left(f^{(r)}, \frac{1}{n} \right) \left(\sum_{j=h}^{v} s_j \lambda_j n^{j-r} \right) = - \frac{\rho_n}{h!} {}^{k_h} D_{1-}^{\alpha_h} x^h + \rho_n =: \psi \tag{2.89}$$

(if $h = 0$, then $\alpha_h = 0$, and $\psi = 0$).

If $h \neq 0$, then

$$\psi = \rho_n \left[1 - \frac{{}^{k_h} D_{1-}^{\alpha_h} x^h}{h!} \right] = \rho_n \left[1 - \int_x^1 k_h (t - x)\, dt \right] =$$

$$\rho_n \left[1 - \int_0^{1-x} k_h(z)\,dz\right] \leq \rho_n \left[1 - \int_0^1 k_h(z)\,dz\right] \leq 0.$$

Hence again in both cases it holds

$$L^*(Q_n(x)) \geq 0,\ \forall\, x \in [-1, 0]. \tag{2.90}$$

Comment 2.12 *Clearly Theorem 2.10 generalizes Theorem 2.5, and Theorem 2.11 generalizes Theorem 4.4, p. 35 of [3]. Furthermore there, the approximating polynomial* Q_n *depends on* f, ρ_n, h; *which* ρ_n *depends on* $n, C_{r,s}, n, v, s_j, \lambda_j$; *which* λ_j *depends on* k_j. *I.e. polynomial* Q_n *among others depends on the type of fractional calculus we use.*

Consequently, Theorem 2.10 is valid for the following left fractional linear differential operators:

(1)

$$L_1^* := \sum_{j=h}^{v} \alpha_j(x) \left[{}^{C}D^{\gamma}_{\rho,\alpha_j,\omega,-1+}\right], \tag{2.91}$$

where $\rho > 0$, $\gamma < 0$, *and* $\omega > 0$ *large enough (from Prabhakar fractional calculus, see (2.26));*

(2)

$$L_2^* := \sum_{j=h}^{v} \alpha_j(x) \left[D^{\alpha_j}_{-1*}\right], \tag{2.92}$$

(see (2.33)) where $\gamma_j > 0$, $j = 1, \ldots, m$; $\lambda = 1$; *and small enough* $\omega_j < 0$, $j = 1, \ldots, m$ *(from generalized non-singular fractional calculus);*

and

(3)

$$L_3^* := \sum_{j=h}^{v} \alpha_j(x) \left[{}^{CF}_{\ \omega}D^{\alpha_j}_{-1+}\right], \tag{2.93}$$

with $\omega < 0$, *sufficiently small (from parametrized Caputo-Fabrizio non-singular kernel fractional calculus).*

Similarly, Theorem 2.11 is valid for the following right fractional linear differential operators:

*(1)**

$$L_1^{**} := \sum_{j=h}^{v} \alpha_j(x) \left[{}^{C}D^{\gamma}_{\rho,\alpha_j,\omega,1-}\right], \tag{2.94}$$

where $\rho > 0$, $\gamma < 0$, *and* $\omega > 0$ *large enough (from Prabhakar fractional calculus, see (2.27));*

*(2)**

$$L_2^{**} := \sum_{j=h}^{v} \alpha_j(x) \left[D^{\alpha_j}_{1-}\right], \tag{2.95}$$

(see (2.34)) where $\gamma_j > 0,\ j = 1, ..., m;\ \lambda = 1$*; and small enough* $\omega_j < 0,\ j = 1, ..., m$ *(from generalized non-singular fractional calculus);*

and

*(3)**

$$L_3^{**} := \sum_{j=h}^{v} \alpha_j(x) \left[{}_{\omega}^{CF} D_{1-}^{\alpha_j} \right], \tag{2.96}$$

with $\omega < 0$*, sufficiently small (from parametrized Caputo-Fabrizio non-singular kernel fractional calculus).*

Our developed abstract simultaneous fractional monotone approximation theory with its applications involves weaker conditions than the one with ordinary derivatives ([2]), and can cover many diverse general cases in a multitude of complex settings and environments.

References

1. Anastassiou, G.A.: Bivariate monotone approximation. Proc. Amer. Math. Soc. **112**(4), 959–964 (1991)
2. Anastassiou, G.A.: High order monotone approximation with linear differential operators. Indian J. Pure Appl. Math. **24**(4), 263–266 (1993)
3. Anastassiou, G.A.: Frontiers in Approximation Theory. World Scientific Publishing Co Pte Ltd., New Jersey, Singapore (2015)
4. Anastassiou, G.A.: Foundations of Generalized Prabhakar-Hilfer Fractional Calculus with Applications. Cubo (2021, accepted)
5. Anastassiou, G.A.: Multiparameter fractional differentiation with non singular kernel. Issues Anal. (2021, accepted)
6. Anastassiou, G.A.: Univariate simultaneous high order abstract fractional monotone approximation with applications (2021, submitted)
7. Anastassiou, G.A., Shisha, O.: Monotone approximation with linear differential operators. J. Approx. Theor. **44**, 391–393 (1985)
8. Atangana, A., Baleanu, D.: New fractional derivatives with nonlocal and non-singular kernel: theory and application to heat transfer model. Therm. Sci. **20**(2), 763–769 (2016)
9. Diethelm, K.: The Analysis of Fractional Differential Equations, Lecture Notes in Mathematics, vol. 2004, 1st ed. Springer, New York, Heidelberg (2010)
10. Giusti, A., et al.: A practical guide to Prabhakar fractional calculus. Fract. Calc. Appl. Anal. **23**(1), 9–54 (2020)
11. Gonska, H.H., Hinnermann, E.: Pointwise estimated for approximation by algebraic polynomials. Acta Math. Hungar. **46**, 243–254 (1985)
12. Gorenflo, R., Kilbas, A., Mainardi, F., Rogosin, S.: Mittag-Leffler functions. Relat. Top. Appl., Springer, Heidelberg, New York (2014)
13. Hewitt, E., Stromberg, K.: Real and Abstract Analysis. Springer, New York, Heidelberg, Berlin (1965)
14. Losada, J., Nieto, J.J.: Properties of a new fractional derivative without singular kernel. Progr. Fract. Differ. Appl. **1** (2), 87–92 (2015)
15. Polito, F., Tomovski, Z.: Some properties of Prabhakar-type fractional calculus operators. Fract. Differ. Calc. **6**(1), 73–94 (2016)
16. Saxena, R.K., Kalla, S.L., Saxena, R.: Integr. Transforms Spec. Functi. Multivariate analogue of generalized Mittag-Leffler function **22**(7), 533–548 (2011)

17. Shisha, O.: Monotone approximation. Pacific J. Math. **15**, 667–671 (1965)
18. Srivastava, H.M., Daoust, M.C.: A note on the convergence of Kompe' de Feriet's double hypergeometric series. Math. Nachr. **53**, 151–159 (1972)

Chapter 3
Spline Abstract Fractional Monotone Approximation

Here we extend our earlier high order simultaneous fractional polynomial spline monotone approximation theory to abstract high order simultaneous fractional polynomial spline monotone approximation, with applications to Prabhakar fractional calculus and non-singular kernel fractional calculi. We cover both the left and right sides of this constrained approximation. Let $f \in C^s([-1, 1])$, $s \in \mathbb{N}$ and L^* be a linear left or right side fractional differential operator such that $L^*(f) \geq 0$ over $[0, 1]$ or $[-1, 0]$, respectively. Then there exists a sequence Q_n, $n \in \mathbb{N}$ of polynomial splines with equally spaced knots of given fixed order such that $L^*(Q_n) \geq 0$ on $[0, 1]$ or $[-1, 0]$, respectively. Furthermore f is approximated with rates fractionally and simultaneously by Q_n in th uniform norm. This constrained fractional approximation on $[-1, 1]$ is given via inequalities involving a higher modulus of smoothness of $f^{(s)}$. It follows [5].

3.1 Introduction

Let $[a, b] \subset \mathbb{R}$ and for $n \geq 1$ consider the partition $\Delta_n := \Delta_n([a, b])$ with points $x_{in} = a + i\left(\frac{b-a}{n}\right)$, $i = 0, 1, ..., n$. Hence $\overline{\Delta_n} \equiv \max_{1 \leq i \leq n}\left(x_{in} - x_{i-1,n}\right) = \frac{b-a}{n}$.

Let $S_m(\Delta_n)$ be the space of polynomial splines of order $m > 0$ with simple knots at the points x_{in}, $i = 1, ..., n-1$ (see [15, p. 5]). Then there exists a linear operator $Q_n : Q_n \equiv Q_n(f)$, mapping $B[a, b]$: the space of bounded real valued functions f on $[a, b]$, into $S_m(\Delta_n)$ (see [15, p. 224], Theorem 6.18).

From the same reference [15, p. 227], Corollary 6.21, we get

G. A. Anastassiou, *Abstract Fractional Monotone Approximation, Theory and Applications*, Studies in Systems, Decision and Control 411,
https://doi.org/10.1007/978-3-030-95943-2_3

Corollary 3.1 *Let $1 \leq \sigma \leq m$, $n \geq 1$. Then for all $f \in C^{\sigma-1}([a, b])$; $r = 0, ..., \sigma - 1$,*

$$\|f^{(r)} - Q_n^{(r)}\| \leq C_1 \left(\frac{b-a}{n}\right)^{\sigma-r-1} \omega_{m-\sigma+1}\left(f^{(\sigma-1)}, \frac{b-a}{n}\right), \tag{3.1}$$

where C_1 depends only on m, $C_1 = C_1(m)$.

By denoting $C_2 = C_1 \max_{0 \leq r \leq \sigma-1} (b-a)^{\sigma-r-1}$ we obtain

Lemma 3.2 ([1]) *Let $1 \leq \sigma \leq m$, $n \geq 1$. Then for all $f \in C^{\sigma-1}([a, b])$; $r = 0, ..., \sigma - 1$,*

$$\|f^{(r)} - Q_n^{(r)}\|_\infty \leq \frac{C_2}{n^{\sigma-r-1}} \omega_{m-\sigma+1}\left(f^{(\sigma-1)}, \frac{b-a}{n}\right), \tag{3.2}$$

where C_2 depends only on m, σ and $b - a$. Here $\omega_{m-\sigma+1}$ is the usual modulus of smoothness of order $m - \sigma + 1$.

We are motivated by

Theorem 3.3 ([1]) *Let h, k, σ, m be integers, $0 \leq h \leq k \leq \sigma - 1$, $\sigma \leq m$ and let $f \in C^{\sigma-1}[a, b]$. Let $\alpha_j(x) \in B[a, b]$, $j = h, h+1, ..., k$ and suppose that $\alpha_h(x) \geq \alpha > 0$ or $\alpha_h(x) \leq \beta < 0$ for all $x \in [a, b]$. Take the linear differential operator*

$$L = \sum_{j=h}^{k} \alpha_j(x) \left[\frac{d^j}{dx^j}\right] \tag{3.3}$$

and assume, throughout $[a, b]$,

$$L(f) \geq 0. \tag{3.4}$$

Then, for every integer $n \geq 1$, there is a polynomial spline function $Q_n(x)$ of order m with simple knots at $\left\{a + i\left(\frac{b-a}{n}\right), i = 1, ..., n-1\right\}$ such that $L(Q_n) \geq 0$ throughout $[a, b]$, and

$$\|f^{(r)} - Q_n^{(r)}\|_\infty \leq \frac{C}{n^{\sigma-k-1}} \omega_{m-\sigma+1}\left(f^{(\sigma-1)}, \frac{b-a}{n}\right), \quad 0 \leq r \leq h. \tag{3.5}$$

Moreover, we find

$$\|f^{(r)} - Q_n^{(r)}\|_\infty \leq \frac{C}{n^{\sigma-r-1}} \omega_{m-\sigma+1}\left(f^{(\sigma-1)}, \frac{b-a}{n}\right), \quad h+1 \leq r \leq \sigma - 1, \tag{3.6}$$

where C is a constant independent of f and n. It depends only on m, σ, L, a, b.

Next we specialize on the case of $a = -1, b = 1$. That is working on $[-1, 1]$. By Lemma 3.2 we get

Lemma 3.4 *Let $1 \leq \sigma \leq m$, $n \geq 1$. Then for all $f \in C^{\sigma-1}([-1, 1])$; $j = 0, 1, ..., \sigma - 1$,*

$$\left\| f^{(j)} - Q_n^{(j)} \right\|_\infty \leq \frac{C_2}{n^{\sigma-j-1}} \omega_{m-\sigma+1}\left(f^{(\sigma-1)}, \frac{2}{n}\right), \tag{3.7}$$

where $C_2 := C_2(m, \sigma) := C_1(m) 2^{\sigma-1}$.

Since

$$\omega_{m-\sigma+1}\left(f^{(\sigma-1)}, \frac{2}{n}\right) \leq 2^{m-\sigma+1} \omega_{m-\sigma+1}\left(f^{(\sigma-1)}, \frac{1}{n}\right) \tag{3.8}$$

(see [7, p. 45]), we obtain

Lemma 3.5 *Let $1 \leq \sigma \leq m$, $n \geq 1$. Then, for each $f \in C^{\sigma-1}([-1, 1])$; $j = 0, 1, ..., \sigma - 1$, there exists $Q_n := Q_n(f) \in S_m(\Delta_n([-1, 1]))$ such that*

$$\left\| f^{(j)} - Q_n^{(j)} \right\|_\infty \leq \frac{C_2^*}{n^{\sigma-j-1}} \omega_{m-\sigma+1}\left(f^{(\sigma-1)}, \frac{1}{n}\right), \tag{3.9}$$

where $C_2^* := C_2^*(m, \sigma) := C_1(m) 2^m$.

We use a lot Lemma 3.5 in this chapter.

In [2], chap. 5, we extended Theorem 3.3 over $[-1, 1]$ to the fractional level. Indeed there L is replaced by L^*, a linear left Caputo fractional differential operator. Then the monotonicity property is only true on the critical interval $[0, 1]$. Simultaneous fractional convergence remained true on all of $[-1, 1]$.

We make

Definition 3.6 ([8, p. 50]) Let $\alpha > 0$ and $\lceil \alpha \rceil = m$, ($\lceil \cdot \rceil$ ceiling of the number). Consider $f \in C^m([-1, 1])$. We define the left Caputo fractional derivative of f of order α as follows:

$$\left(D_{*-1}^{\alpha} f\right)(x) = \frac{1}{\Gamma(m-\alpha)} \int_{-1}^{x} (x-t)^{m-\alpha-1} f^{(m)}(t)\, dt, \tag{3.10}$$

for any $x \in [-1, 1]$, where Γ is the gamma function.

We set

$$\begin{aligned} D_{*-1}^{0} f(x) &= f(x), \\ D_{*-1}^{m} f(x) &= f^{(m)}(x), \ \forall\, x \in [-1, 1]. \end{aligned} \tag{3.11}$$

So in [2], chap. 5, we proved

Theorem 3.7 *Let h, k, σ, m be integers, $1 \leq \sigma \leq m$, $n \in \mathbb{N}$, with $0 \leq h \leq k \leq \sigma - 2$ and let $f \in C^{\sigma-1}([-1, 1])$, with $f^{(\sigma-1)}$ having modulus of smoothness $\omega_{m-\sigma+1}\left(f^{(\sigma-1)}, \delta\right)$ there, $\delta > 0$. Let $\alpha_j(x)$, $j = h, h+1, ..., k$ be real functions, defined and bounded on $[-1, 1]$ and suppose $\alpha_h(x)$ is either $\geq \alpha > 0$ or $\leq \beta < 0$ on $[0, 1]$. Let the real numbers $\alpha_0 = 0 < \alpha_1 \leq 1 < \alpha_2 \leq 2 < ... < \alpha_{\sigma-2} \leq \sigma - 2$. Here $D_{*-1}^{\alpha_j} f$ stands for the left Caputo fractional derivative of f of order α_j anchored at -1. Consider the linear left fractional differential operator*

$$L^* := \sum_{j=h}^{k} \alpha_j(x)\left[D_{*-1}^{\alpha_j}\right] \tag{3.12}$$

and suppose, throughout $[0, 1]$, $L^*(f) \geq 0$.

Then, for every integer $n \geq 1$, there exists a polynomial spline function $Q_n(x)$ of order $m > 0$ with simple knots at $\left\{-1 + i\frac{2}{n},\ i = 1, ..., n-1\right\}$ such that $L^*(Q_n) \geq 0$ throughout $[0, 1]$, and

$$\sup_{-1 \leq x \leq 1} \left|\left(D_{*-1}^{\alpha_j} f\right)(x) - \left(D_{*-1}^{\alpha_j} Q_n\right)(x)\right| \leq$$

$$\frac{2^{j-\alpha_j}}{\Gamma\left(j - \alpha_j + 1\right)} \frac{C_2^*}{n^{\sigma-j-1}} \omega_{m-\sigma+1}\left(f^{(\sigma-1)}, \frac{1}{n}\right), \tag{3.13}$$

$j = h+1, ..., \sigma - 2$.

Set

$$l_j :\equiv \sup_{x\in[-1,1]} \left|\alpha_h^{-1}(x)\, \alpha_j(x)\right|,\ h \leq j \leq k. \tag{3.14}$$

When $j = 1, ..., h$ we derive

$$\max_{-1\leq x\leq 1} \left|\left(D_{*-1}^{\alpha_j} f\right)(x) - \left(D_{*-1}^{\alpha_j} Q_n\right)(x)\right| \leq \frac{C_2^*}{n^{\sigma-k-1}} \omega_{m-\sigma+1}\left(f^{(\sigma-1)}, \frac{1}{n}\right) \cdot$$

$$\left[\left(\sum_{\tau=h}^{k} l_\tau \frac{2^{\tau-\alpha_\tau}}{\Gamma\left(\tau - \alpha_\tau + 1\right)}\right)\left(\sum_{\lambda=0}^{h-j} \frac{2^{h-\alpha_j-\lambda}}{\lambda!\Gamma\left(h - \alpha_j - \lambda + 1\right)}\right) + \frac{2^{j-\alpha_j}}{\Gamma\left(j - \alpha_j + 1\right)}\right]. \tag{3.15}$$

Finally it holds

$$\sup_{-1\leq x\leq 1} \left|f(x) - Q_n(x)\right| \leq$$

$$\frac{C_2^*}{n^{\sigma-k-1}} \omega_{m-\sigma+1}\left(f^{(\sigma-1)}, \frac{1}{n}\right)\left[\frac{1}{h!}\sum_{\tau=h}^{k} l_\tau \frac{2^{\tau-\alpha_\tau}}{\Gamma\left(\tau - \alpha_\tau + 1\right)} + 1\right]. \tag{3.16}$$

In this chapter we establish abstract fractional calculus left and right high order monotone approximation theory (by polynomial splines of simple knots) of Caputo type, and then we apply our results to Prabhakar fractional Calculus, generalized non-singular fractional calculus, and parametrized Caputo-Fabrizio non-singular fractional calculus.

Next, we build the related necessary fractional calculi background.

3.2 Fractional Calculi

Here we need to be very specific in preparation for our main results.

3.2.1 *Abstract Fractional Calculus*

Let h, v, σ, m be non-negative integers, $2 \leq \sigma \leq m$, with $0 \leq h \leq v \leq \sigma - 2$. Let also $\mathbb{N} \not\ni \alpha_j > 0$, $j = 1, ..., \sigma - 2$, such that $\alpha_0 = 0 < \alpha_1 < 1 < \alpha_2 < 2 < \alpha_3 < 3 < ... < ... < \alpha_{\sigma-2} < \sigma - 2$. That is $\lceil \alpha_j \rceil = j$, $j = 1, ..., \sigma - 2$; $\lceil \alpha_0 \rceil = 0$.

Consider the integrable functions $k_j := K_{\alpha_j} : [0, 2] \to \mathbb{R}_+$, $j = 0, 1, ..., \sigma - 2$. Here $g \in C^{\sigma-1}([-1, 1])$.

We consider the following abstract left side Caputo type fractional derivatives:

$$\left({}^{k_j}D_{*-1}^{\alpha_j} g\right)(x) := \int_{-1}^{x} k_j(x - t)\, g^{(j)}(t)\, dt, \tag{3.17}$$

$j = 1, .., \sigma - 2$; $\forall\, x \in [-1, 1]$.

Similarly, we define the corresponding right side generalized Caputo type fractional derivatives:

$$\left({}^{k_j}D_{1-}^{\alpha_j} g\right)(x) := (-1)^j \int_{x}^{1} k_j(t - x)\, g^{(j)}(t)\, dt, \tag{3.18}$$

$j = 1, .., \sigma - 2$; $\forall\, x \in [-1, 1]$.

We set

$$\left({}^{k_j}D_{*-1}^{j} g\right)(x) := g^{(j)}(x); \quad \left({}^{k_j}D_{1-}^{j} g\right)(x) := (-1)^j g^{(j)}(x), \tag{3.19}$$

for $j = 1, ..., \sigma - 2$, and also we set

$$\left({}^{k_0}D_{*-1}^{0} g\right)(x) := \left({}^{k_0}D_{1-}^{0} g\right)(x) := g(x), \tag{3.20}$$

$\forall\, x \in [-1, 1]$.

We will assume that

$$\int_0^1 k_h(z)\,dz \geq 1, \text{ when } h \neq 0. \tag{3.21}$$

In the usual Caputo fractional derivatives case it is

$$k_j(z) = \frac{z^{j-\alpha_j-1}}{\Gamma\left(j-\alpha_j\right)}, \ j = 1, ..., \sigma - 2; \ \forall z \in [0, 2], \tag{3.22}$$

and (3.21) is fulfilled, by the fact that $\Gamma(h - \alpha_h + 1) \leq 1$, see [2, p. 6].

3.2.2 *About Prabhakar Fractional Calculus*

Here we follow [3, 13].

We consider the Prabhakar function (also known as the three parameter Mittag-Leffler function), (see [10, p. 97]; [9])

$$E^{\gamma}_{\alpha,\beta}(z) = \sum_{k=0}^{\infty} \frac{(\gamma)_k}{k!\Gamma(\alpha k + \beta)} z^k, \tag{3.23}$$

where Γ is the gamma function; $\alpha, \beta > 0$, $\gamma \in \mathbb{R}$, $z \in \mathbb{R}$, and $(\gamma)_k = \gamma(\gamma+1)\ldots(\gamma+k-1)$. It is $E^0_{\alpha,\beta}(z) = \frac{1}{\Gamma(\beta)}$.

Let $a, b \in \mathbb{R}$, $a < b$ and $x \in [a, b]$; $f \in C([a, b])$. The left and right Prabhakar fractional integrals are defined ([3, 13]) as follows:

$$\left(e^{\gamma}_{\rho,\mu,\omega,a+}f\right)(x) = \int_a^x (x-t)^{\mu-1} E^{\gamma}_{\rho,\mu}\left[\omega(x-t)^{\rho}\right] f(t)\,dt, \tag{3.24}$$

and

$$\left(e^{\gamma}_{\rho,\mu,\omega,b-}f\right)(x) = \int_x^b (t-x)^{\mu-1} E^{\gamma}_{\rho,\mu}\left[\omega(t-x)^{\rho}\right] f(t)\,dt, \tag{3.25}$$

where $\rho, \mu > 0$; $\gamma, \omega \in \mathbb{R}$.

Functions (3.24) and (3.25) are continuous, see [3].

Next, let $f \in C^N([a, b])$, where $N = \lceil \mu \rceil$, ($\lceil \cdot \rceil$ is the ceiling of the number), $0 < \mu \notin \mathbb{N}$. We define the Prabhakar-Caputo left and right fractional derivatives of order μ ([3, 13]) as follows ($x \in [a, b]$):

$$\left({}^{C}D^{\gamma}_{\rho,\mu,\omega,a+}f\right)(x) = \int_{a}^{x} (x-t)^{N-\mu-1} E^{-\gamma}_{\rho,N-\mu}\left[\omega (x-t)^{\rho}\right] f^{(N)}(t)\,dt, \quad (3.26)$$

with ${}^{C}D^{\gamma}_{\rho,0,\omega,a+}f := f$; ${}^{C}D^{\gamma}_{\rho,N,\omega,a+}f := f^{(N)}$, $N \in \mathbb{N}$,

and

$$\left({}^{C}D^{\gamma}_{\rho,\mu,\omega,b-}f\right)(x) = (-1)^{N} \int_{x}^{b} (t-x)^{N-\mu-1} E^{-\gamma}_{\rho,N-\mu}\left[\omega (t-x)^{\rho}\right] f^{(N)}(t)\,dt, \quad (3.27)$$

with ${}^{C}D^{\gamma}_{\rho,0,\omega,b-}f := f$; ${}^{C}D^{\gamma}_{\rho,N,\omega,b-}f := (-1)^{N} f^{(N)}$, $N \in \mathbb{N}$.

One can rewrite (3.26) and (3.27) as follows:

$$\left({}^{C}D^{\gamma}_{\rho,\mu,\omega,a+}f\right)(x) = \left(e^{-\gamma}_{\rho,N-\mu,\omega,a+}f^{[N]}\right)(x), \quad (3.28)$$

and

$$\left({}^{C}D^{\gamma}_{\rho,\mu,\omega,b-}f\right)(x) = (-1)^{N}\left(e^{-\gamma}_{\rho,N-\mu,\omega,b-}f^{[N]}\right)(x), \quad (3.29)$$

$\forall\, x \in [a, b]$.

Clearly, the functions (3.28) and (3.29) are continuous.

3.2.3 *From Generalized Non-singular Fractional Calculus ([4])*

We need

Definition 3.8 Here we use the multivariate analogue of generalized Mittag-Leffler function, see [14], defined for $\lambda, \gamma_j, \rho_j, z_j \in \mathbb{C}$, $Re\left(\rho_j\right) > 0\,(j = 1, ..., m)$ in terms of a multiple series of the form:

$$E^{(\gamma_j)}_{(\rho_j),\lambda}(z_1, ..., z_m) = E^{(\gamma_1,...,\gamma_m)}_{(\rho_1,...,\rho_m),\lambda}(z_1, ..., z_m) =$$

$$\sum_{k_1,...,k_m=0}^{\infty} \frac{(\gamma_1)_{k_1} \cdots (\gamma_m)_{k_m}}{\Gamma\left(\lambda + \sum_{j=1}^{m} k_j \rho_j\right)} \frac{z_1^{k_1} ... z_m^{k_m}}{k_1! ... k_m!}, \quad (3.30)$$

where $\left(\gamma_j\right)_{k_j}$ is the Pochhammer symbol, Γ is the gamma function. By [16, p. 157], (3.30) converges for $Re\left(\rho_j\right) > 0$, $j = 1, ..., m$.

In what follows we will use the particular case of $E_{(\rho,\ldots,\rho),\lambda}^{(\gamma_1,\ldots,\gamma_m)}[\omega_1 t^\rho, \ldots, \omega_m t^\rho]$, denoted by $E_{(\rho),\lambda}^{(\gamma_j)}[\omega_1 t^\rho, \ldots, \omega_m t^\rho]$, where $0 < \rho < 1$, $t \geq 0$, $\lambda > 0$, $\gamma_j \in \mathbb{R}$ with $(\gamma_j)_{k_j} := \gamma_j (\gamma_j + 1) \ldots (\gamma_j + k_j - 1)$, $\omega_j \in \mathbb{R} - \{0\}$, for $j = 1, \ldots, m$.

Let now $f \in C^{n+1}([a, b])$, $n \in \mathbb{Z}_+$.

We define the Caputo type generalized left fractional derivative with non-singular kernel of order $n + \rho$, as

$$D_{a*}^{n+\rho} f(x) := {}_{(\gamma_j)(\omega_j)}^{CA} D_{a*}^{n+\rho,\lambda} f(x) :=$$

$$\frac{A(\rho)}{1-\rho} \int_a^x E_{(\rho),\lambda}^{(\gamma_j)} \left[\frac{-\omega_1 \rho}{1-\rho} (x-t)^\rho, \ldots, \frac{-\omega_m \rho}{1-\rho} (x-t)^\rho \right] f^{(n+1)}(t)\, dt, \tag{3.31}$$

$\forall\, x \in [a, b]$.

Similarly, we define the Caputo type generalized right fractional derivative with non-singular kernel of order $n + \rho$, as

$$D_{b-}^{n+\rho} f(x) := {}_{(\gamma_j)(\omega_j)}^{CA} D_{b-}^{n+\rho,\lambda} f(x) :=$$

$$(-1)^{n+1} \frac{A(\rho)}{1-\rho} \int_x^b E_{(\rho),\lambda}^{(\gamma_j)} \left[\frac{-\omega_1 \rho}{1-\rho} (t-x)^\rho, \ldots, \frac{-\omega_m \rho}{1-\rho} (t-x)^\rho \right] f^{(n+1)}(t)\, dt, \tag{3.32}$$

$\forall\, x \in [a, b]$.

Above $A(\rho)$ is a normalizing constant.

The above derivatives (3.31), (3.32) generalize the Atangana-Baleanu fractional derivatives [6].

We rewrite (3.31) and (3.32), and for $[a, b] = [-1, 1]$.

Let $\mu > 0$ with $\mu \notin \mathbb{N}$ and $\lceil \mu \rceil = n \in \mathbb{N}$. That is $0 < 1 - n + \mu < 1$, and let $f \in C^n([-1, 1])$. Then, we have

$$D_{-1*}^{\mu} f(x) := {}_{(\gamma_j)(\omega_j)}^{CA} D_{-1*}^{\mu,\lambda} f(x) :=$$

$$\frac{A(1-n+\mu)}{n-\mu} \int_{-1}^x E_{(1-n+\mu),\lambda}^{(\gamma_j)} \left[\frac{-\omega_1 (1-n+\mu)}{n-\mu} (x-t)^{1-n+\mu}, \ldots, \right. \tag{3.33}$$

$$\left. \frac{-\omega_m (1-n+\mu)}{n-\mu} (x-t)^{1-n+\mu} \right] f^{(n)}(t)\, dt,$$

and

$$D_{1-}^{\mu} f(x) := {}_{(\gamma_j)(\omega_j)}^{CA} D_{1-}^{\mu,\lambda} f(x) :=$$

$$(-1)^n \frac{A(1-n+\mu)}{n-\mu} \int_x^1 E^{(\gamma_j)}_{(1-n+\mu),\lambda} \left[\frac{-\omega_1 (1-n+\mu)}{n-\mu} (t-x)^{1-n+\mu}, \ldots, \right. \quad (3.34)$$

$$\left. \frac{-\omega_m (1-n+\mu)}{n-\mu} (t-x)^{1-n+\mu} \right] f^{(n)}(t)\, dt,$$

$\forall\, x \in [-1, 1]$.

We will set $D^0_{-1*} f = f$, $D^0_{1-} f = f$, and $D^m_{-1*} f = f^{(m)}$, $D^m_{1-} f = (-1)^m f^{(m)}$, when $m \in \mathbb{N}$.

We make

Remark 3.9 Fractional Calculi of Sects. 3.2.2 and 3.2.3 are special cases of abstract fractional calculus, see Sect. 3.2.1. In particular the important condition (3.21) is fulfilled.

So, we will verify $\int_0^1 k_h(z)\, dz \geq 1$, $h \neq 0$.

(I) First for Sect. 3.2.2:

We notice that

$$\int_0^1 z^{N-\mu-1} E^{-\gamma}_{\rho,N-\mu}(\omega z^\rho)\, dz =$$

(here $\rho, N-\mu > 0$, $\gamma < 0$, $\omega > 0$)

$$\int_0^1 z^{N-\mu-1} \sum_{k=0}^{\infty} \frac{(-\gamma)_k}{k!\Gamma(\rho k + N - \mu)} (\omega z^\rho)^k\, dz \overset{\text{(by [11], p. 175)}}{=}$$

$$\sum_{k=0}^{\infty} \frac{(-\gamma)_k\, \omega^k}{k!\Gamma(\rho k + N - \mu)} \int_0^1 z^{N-\mu-1} z^{\rho k} dz = \quad (3.35)$$

$$\sum_{k=0}^{\infty} \frac{(-\gamma)_k\, \omega^k}{k!\Gamma(\rho k + N - \mu)} \int_0^1 z^{(\rho k+N)-\mu-1} dz =$$

$$\sum_{k=0}^{\infty} \frac{(-\gamma)_k\, \omega^k}{k!\Gamma(\rho k + N - \mu + 1)} = E^{-\gamma}_{\rho,N-\mu+1}(\omega) \geq 1,$$

for suitable $\omega > 0$.

(II) Next, for Sect. 3.2.3:

Here $\gamma_j > 0$, $j = 1, \ldots, m$; $\lambda = 1$; $\mathbb{N} \not\ni \mu > 0$, $\lceil \mu \rceil = n \in \mathbb{N}$, $\omega_j < 0$, $j = 1, \ldots, m$.

Without loss of generality we assume that $A(1-n+\mu) > 0$.

We have that

$$\frac{A(1-n+\mu)}{n-\mu}\int_0^1 E_{(1-n+\mu),1}^{(\gamma_j)}\left[\frac{-\omega_1(1-n+\mu)}{n-\mu}z^{1-n+\mu},\right.$$

$$\left.\ldots,\frac{-\omega_m(1-n+\mu)}{n-\mu}z^{1-n+\mu}\right]dz=$$

(here $0<1-(n-\mu)=1-n+\mu<1$)

$$\frac{A(1-n+\mu)}{n-\mu}\int_0^1\left[\sum_{k_1,\ldots,k_m=0}^{\infty}\frac{(\gamma_1)_{k_1}\ldots(\gamma_m)_{k_m}}{\Gamma\left(1+\left(\sum_{j=1}^m k_j\right)(1-n+\mu)\right)}\right.$$

$$\left.\frac{\prod_{j=1}^m\left(\left(\frac{-\omega_j(1-n+\mu)}{n-\mu}\right)^{k_j}z^{(1-n+\mu)\sum_{j=1}^m k_j}\right)}{k_1!\ldots k_m!}\right]dz$$

$$=\left(\frac{A(1-n+\mu)}{n-\mu}\right)$$

$$\sum_{k_1,\ldots,k_m=0}^{\infty}\frac{(\gamma_1)_{k_1}\ldots(\gamma_m)_{k_m}}{\Gamma\left(1+\left(\sum_{j=1}^m k_j\right)(1-n+\mu)\right)}\frac{\prod_{j=1}^m\left(\frac{-\omega_j(1-n+\mu)}{n-\mu}\right)^{k_j}}{k_1!\ldots k_m!}\int_0^1 z^{(1-n+\mu)\sum_{j=1}^m k_j}dz \tag{3.36}$$

$$=\left(\frac{A(1-n+\mu)}{n-\mu}\right)\sum_{k_1,\ldots,k_m=0}^{\infty}\frac{(\gamma_1)_{k_1}\ldots(\gamma_m)_{k_m}}{\Gamma\left(2+\left(\sum_{j=1}^m k_j\right)(1-n+\mu)\right)}\frac{\prod_{j=1}^m\left(\frac{-\omega_j(1-n+\mu)}{n-\mu}\right)^{k_j}}{k_1!\ldots k_m!}$$

$$=\frac{A(1-n+\mu)}{n-\mu}E_{(1-n+\mu),2}^{(\gamma_j)}\left(\frac{-\omega_1(1-n+\mu)}{n-\mu},\ldots,\frac{-\omega_m(1-n+\mu)}{n-\mu}\right)\geq 1, \tag{3.37}$$

for suitable $\omega_j<0$, for $j=1,\ldots,m$.

We also need

Definition 3.10 Let $f \in C^n([-1,1])$, $\mathbb{N} \not\ni \mu > 0$, $\lceil \mu \rceil = n \in \mathbb{N}$; $\omega < 0$. That is $0 < 1 - n + \mu < 1$. The parametrized Caputo-Fabrizio non-singular kernel fractional derivatives, left and right of order μ, respectively, are given as follows (see also [12]):

$$ {}_{\omega}^{CF}D_{-1+}^{\mu} f(x) := \frac{1}{n-\mu} \int_{-1}^{x} \exp\left(-\frac{(1-n+\mu)\,\omega}{n-\mu} (x-t) \right) f^{(n)}(t)\, dt, \tag{3.38} $$

$$ {}_{\omega}^{CF}D_{1-}^{\mu} f(x) := \frac{(-1)^n}{n-\mu} \int_{x}^{1} \exp\left(-\frac{(1-n+\mu)\,\omega}{n-\mu} (t-x) \right) f^{(n)}(t)\, dt, \tag{3.39} $$

$\forall\, x \in [-1,1]$.

Equations (3.38), (3.39) are special cases of (3.17), (3.18).

We make

Remark 3.11 We want to calculate

$$ \infty > \int_{0}^{1} \exp\left(-\frac{(1-n+\mu)\,\omega}{n-\mu} z \right) dz $$

(set $\delta := -\frac{(1-n+\mu)\omega}{n-\mu}$)

$$ = \int_{0}^{1} e^{\delta z} dz = \frac{1}{\delta} e^{\delta z} \big|_0^1 = \frac{1}{\delta}\left(e^{\delta} - 1\right) = \frac{e^{\delta}}{\delta} - \frac{1}{\delta} = \frac{\sum_{k=0}^{\infty} \frac{\delta^k}{k!}}{\delta} - \frac{1}{\delta} \tag{3.40} $$

$$ = \sum_{k=0}^{\infty} \frac{\delta^{k-1}}{k!} = \sum_{k=0}^{\infty} \left(\frac{\left(\frac{1-n+\mu}{n-\mu}\right)^{k-1} (-\omega)^{k-1}}{k!} \right) \geq 1, $$

for suitable $\omega < 0$.

So, again condition (3.21) is valid.

3.3 Main Results

We present

Theorem 3.12 *Let h, v, σ, m be integers, $2 \leq \sigma \leq m$, $n \in \mathbb{N}$, with $0 \leq h \leq v \leq \sigma - 2$ and let $f \in C^{\sigma-1}([-1, 1])$, with $f^{(\sigma-1)}$ having modulus of smoothness $\omega_{m-\sigma+1}\left(f^{(\sigma-1)}, \delta\right)$ there, $\delta > 0$. Let $\alpha_j(x)$, $j = h, h+1, ..., v$ be real functions, defined and bounded on $[-1, 1]$ and suppose $\alpha_h(x)$ is either $\geq \alpha > 0$ or $\leq \beta < 0$ on $[0, 1]$. Let the real numbers $\alpha_0 = 0 < \alpha_1 < 1 < \alpha_2 < 2 < ... < \alpha_{\sigma-2} < \sigma - 2$. Here ${}^{k_j}D_{*-1}^{\alpha_j} f$ stands for the abstract left Caputo fractional derivative of f of order α_j anchored at -1. We consider the abstract left fractional linear differential operator*

$$L^* := \sum_{j=h}^{v} \alpha_j(x)\left[{}^{k_j}D_{*-1}^{\alpha_j}\right], \tag{3.41}$$

and suppose, throughout $[0, 1]$,

$$L^*(f) \geq 0. \tag{3.42}$$

Set

$$\lambda_j := \int_0^2 k_j(z)\, dz, \; j = 1, ..., \sigma - 2, \tag{3.43}$$

and $\lambda_0 = 1$; along with

$$s_j := \sup_{-1 \leq x \leq 1} \left|\alpha_h^{-1}(x)\, \alpha_j(x)\right|, \; j = h, ..., u. \tag{3.44}$$

Then, for any $n \in \mathbb{N}$ there exists a spline $Q_n(x) \in S_m(\Delta_n([-1, 1]))$ such that

$$L^*(Q_n) \geq 0, \text{ throughout } [0, 1], \tag{3.45}$$

and
(i)

$$\max_{-1 \leq x \leq 1} \left|\left({}^{k_j}D_{*-1}^{\alpha_j} f\right)(x) - \left({}^{k_j}D_{*-1}^{\alpha_j} Q_n\right)(x)\right| \leq \lambda_j \frac{C_2^*}{n^{\sigma-j-1}} \omega_{m-\sigma+1}\left(f^{(\sigma-1)}, \frac{1}{n}\right), \tag{3.46}$$

when $j = h+1, ..., \sigma - 2$;
(ii) *if $h \neq 0$ and $j = 1, ..., h$, we have*

$$\max_{-1 \leq x \leq 1} \left|\left({}^{k_j}D_{*-1}^{\alpha_j} f\right)(x) - \left({}^{k_j}D_{*-1}^{\alpha_j} Q_n\right)(x)\right| \leq$$

$$\frac{C_2^* \omega_{m-\sigma+1}\left(f^{(\sigma-1)}, \frac{1}{n}\right)}{n^{\sigma-1-v}} \left[\frac{\left(\sum_{j_*=h}^{v} s_{j_*} \lambda_{j_*}\right)}{h!} \max_{-1 \leq x \leq 1} \left| \left({}^{k_j} D_{*-1}^{\alpha_j} x^h \right) \right| + \lambda_j \right]; \tag{3.47}$$

(iii) *furthermore*

$$\max_{-1 \leq x \leq 1} |f(x) - Q_n(x)| \leq \frac{C_2^* \omega_{m-\sigma+1}\left(f^{(\sigma-1)}, \frac{1}{n}\right)}{n^{\sigma-1-v}} \left[\frac{\left(\sum_{j=h}^{v} s_j \lambda_j\right)}{h!} + 1 \right]. \tag{3.48}$$

Proof Let Q_n be as in Lemma 3.5. Let also $\alpha_j > 0$, $j = 1, ..., \sigma - 2$, such that $\alpha_0 = 0 < \alpha_1 < 1 < \alpha_2 < 2 < ... < \alpha_{\sigma-2} < \sigma - 2$. That is $\lceil \alpha_j \rceil = j$, $j = 0, 1, ..., \sigma - 2$.

We consider the abstract left Caputo fractional derivatives ${}^{k_j} D_{*-1}^{\alpha_j} f$ and ${}^{k_j} D_{*-1}^{\alpha_j} Q_n$, $j = 1, ..., \sigma - 2$.

We notice that ($x \in [-1, 1]$)

$$\left| \left({}^{k_j} D_{*-1}^{\alpha_j} f \right)(x) - \left({}^{k_j} D_{*-1}^{\alpha_j} Q_n \right)(x) \right| =$$

$$\left| \int_{-1}^{x} k_j(x-t) f^{(j)}(t)\, dt - \int_{-1}^{x} k_j(x-t) Q_n^{(j)}(t)\, dt \right| = \tag{3.49}$$

$$\left| \int_{-1}^{x} k_j(x-t) \left(f^{(j)}(t) - Q_n^{(j)}(t) \right) dt \right| \leq$$

$$\int_{-1}^{x} k_j(x-t) \left| f^{(j)}(t) - Q_n^{(j)}(t) \right| dt \overset{(3.9)}{\leq}$$

$$\left(\int_{-1}^{x} k_j(x-t)\, dt \right) \frac{C_2^*}{n^{\sigma-j-1}} \omega_{m-\sigma+1}\left(f^{(\sigma-1)}, \frac{1}{n} \right) =$$

$$\left(\int_{0}^{x+1} k_j(z)\, dz \right) \frac{C_2^*}{n^{\sigma-j-1}} \omega_{m-\sigma+1}\left(f^{(\sigma-1)}, \frac{1}{n} \right) \leq$$

$$\left(\int_0^2 k_j(z)\,dz\right)\frac{C_2^*}{n^{\sigma-j-1}}\omega_{m-\sigma+1}\left(f^{(\sigma-1)},\frac{1}{n}\right). \tag{3.50}$$

We have proved that

$$\left|\left({}^{k_j}D_{*-1}^{\alpha_j}f\right)(x)-\left({}^{k_j}D_{*-1}^{\alpha_j}Q_n\right)(x)\right|\le$$

$$\left(\int_0^2 k_j(z)\,dz\right)\frac{C_2^*}{n^{\sigma-j-1}}\omega_{m-\sigma+1}\left(f^{(\sigma-1)},\frac{1}{n}\right), \tag{3.51}$$

$\forall\, x\in[-1,1]$, $j=1,...,\sigma-2$, $n\ge1$.

Hence it holds

$$\max_{-1\le x\le1}\left|\left({}^{k_j}D_{*-1}^{\alpha_j}f\right)(x)-\left({}^{k_j}D_{*-1}^{\alpha_j}Q_n\right)(x)\right|\le \tag{3.52}$$

$$\lambda_j\frac{C_2^*}{n^{\sigma-j-1}}\omega_{m-\sigma+1}\left(f^{(\sigma-1)},\frac{1}{n}\right),\ j=1,...,\sigma-2,\ n\ge1,$$

where

$$\lambda_j:=\int_0^2 k_j(z)\,dz,\ j=1,...,\sigma-2,\ \text{as in (3.43.).}$$

Inequality (3.52) is valid when $j=0$ by (3.9) and (3.20), and we can set $\lambda_0=1$.

Set

$$\gamma_n:=C_2^*\omega_{m-\sigma+1}\left(f^{(\sigma-1)},\frac{1}{n}\right)\left(\sum_{j=h}^{v}s_j\lambda_j n^{j-\sigma+1}\right). \tag{3.53}$$

I. Suppose, throughout [0, 1], $\alpha_h(x)\ge\alpha>0$.

Let $Q_n(x)\in S_m(\Delta_n([-1,1]))$, $x\in[-1,1]$, as in Lemma 3.5, so that

$$\max_{-1\le x\le1}\left|{}^{k_j}D_{*-1}^{\alpha_j}\left(f(x)+\gamma_n(h!)^{-1}x^h\right)-\left({}^{k_j}D_{*-1}^{\alpha_j}Q_n\right)(x)\right|\overset{(3.52.)}{\le} \tag{3.54}$$

$$\lambda_j\frac{C_2^*}{n^{\sigma-j-1}}\omega_{m-\sigma+1}\left(f^{(\sigma-1)},\frac{1}{n}\right),\ j=0,1,...,\sigma-2.$$

When $j=h+1,...,\sigma-2$, we get

$$\max_{-1\le x\le1}\left|\left({}^{k_j}D_{*-1}^{\alpha_j}f\right)(x)-\left({}^{k_j}D_{*-1}^{\alpha_j}Q_n\right)(x)\right|\le\lambda_j\frac{C_2^*}{n^{\sigma-j-1}}\omega_{m-\sigma+1}\left(f^{(\sigma-1)},\frac{1}{n}\right), \tag{3.55}$$

proving (3.46).

For $j = 1, ..., h \neq 0$, we have that

$$\max_{-1\leq x\leq 1}\left|\left({}^{k_j}D_{*-1}^{\alpha_j}f\right)(x)-\left({}^{k_j}D_{*-1}^{\alpha_j}Q_n\right)(x)\right|\overset{(3.54.)}{\leq}$$

$$\frac{\gamma_n}{h!}\max_{-1\leq x\leq 1}\left|\left({}^{k_j}D_{*-1}^{\alpha_j}x^h\right)\right|+\lambda_j\frac{C_2^*}{n^{\sigma-j-1}}\omega_{m-\sigma+1}\left(f^{(\sigma-1)},\frac{1}{n}\right)=$$

$$\frac{C_2^*}{h!}\omega_{m-\sigma+1}\left(f^{(\sigma-1)},\frac{1}{n}\right)\left(\sum_{j_*=h}^{v}s_{j_*}\lambda_{j_*}n^{j_*-\sigma+1}\right)\max_{-1\leq x\leq 1}\left|\left({}^{k_j}D_{*-1}^{\alpha_j}x^h\right)\right|$$

$$+\lambda_j\frac{C_2^*}{n^{\sigma-j-1}}\omega_{m-\sigma+1}\left(f^{(\sigma-1)},\frac{1}{n}\right)= \tag{3.56}$$

$$C_2^*\omega_{m-\sigma+1}\left(f^{(\sigma-1)},\frac{1}{n}\right)\left[\frac{1}{h!}\left(\sum_{j_*=h}^{v}s_{j_*}\lambda_{j_*}n^{j_*-\sigma+1}\right)\max_{-1\leq x\leq 1}\left|\left({}^{k_j}D_{*-1}^{\alpha_j}x^h\right)\right|+\frac{\lambda_j}{n^{\sigma-j-1}}\right]\leq$$

$$\frac{C_2^*\omega_{m-\sigma+1}\left(f^{(\sigma-1)},\frac{1}{n}\right)}{n^{\sigma-1-v}}\left[\frac{\left(\sum_{j_*=h}^{v}s_{j_*}\lambda_{j_*}\right)}{h!}\max_{-1\leq x\leq 1}\left|\left({}^{k_j}D_{*-1}^{\alpha_j}x^h\right)\right|+\lambda_j\right],$$

proving (3.47).

When $j = 0$ from (3.54) we get

$$\max_{-1\leq x\leq 1}\left|f(x)+\gamma_n(h!)^{-1}x^h-Q_n(x)\right|\leq\frac{C_2^*}{n^{\sigma-1}}\omega_{m-\sigma+1}\left(f^{(\sigma-1)},\frac{1}{n}\right). \tag{3.57}$$

That is

$$\max_{-1\leq x\leq 1}\left|f(x)-Q_n(x)\right|\leq\frac{\gamma_n}{h!}+\frac{C_2^*}{n^{\sigma-1}}\omega_{m-\sigma+1}\left(f^{(\sigma-1)},\frac{1}{n}\right)= \tag{3.58}$$

$$\frac{C_2^*\omega_{m-\sigma+1}\left(f^{(\sigma-1)},\frac{1}{n}\right)}{h!}\left(\sum_{j=h}^{v}s_j\lambda_jn^{j-\sigma+1}\right)+\frac{C_2^*}{n^{\sigma-1}}\omega_{m-\sigma+1}\left(f^{(\sigma-1)},\frac{1}{n}\right)\leq$$

$$C_2^*\omega_{m-\sigma+1}\left(f^{(\sigma-1)},\frac{1}{n}\right)\left[\frac{1}{h!n^{\sigma-1-v}}\left(\sum_{j=h}^{v}s_j\lambda_j\right)+\frac{1}{n^{\sigma-1}}\right]\leq \tag{3.59}$$

$$\frac{C_2^* \omega_{m-\sigma+1}\left(f^{(\sigma-1)}, \frac{1}{n}\right)}{n^{\sigma-1-v}} \left[\frac{\left(\sum_{j=h}^{v} s_j \lambda_j\right)}{h!} + 1 \right],$$

proving (3.48).

Also if $0 \leq x \leq 1$, then

$$\alpha_h^{-1}(x) L^*(Q_n(x)) = \alpha_h^{-1}(x) L^*(f(x)) + \frac{\gamma_n}{h!} {}^{k_h}D_{*-1}^{\alpha_h} x^h + \tag{3.60}$$

$$\sum_{j=h}^{v} \alpha_h^{-1}(x) \alpha_j(x) \left[{}^{k_j}D_{*-1}^{\alpha_j} Q_n(x) - {}^{k_j}D_{*-1}^{\alpha_j} f(x) - \frac{\gamma_n}{h!} {}^{k_j}D_{*-1}^{\alpha_j} x^h \right] \overset{(3.54.)}{\geq}$$

(by $L^* f \geq 0$)

$$\frac{\gamma_n}{h!} {}^{k_h}D_{*-1}^{\alpha_h} x^h - C_2^* \omega_{m-\sigma+1}\left(f^{(\sigma-1)}, \frac{1}{n}\right) \left(\sum_{j=h}^{v} s_j \lambda_j n^{j-\sigma+1} \right) =$$

$$\frac{\gamma_n}{h!} {}^{k_h}D_{*-1}^{\alpha_h} x^h - \gamma_n =: \xi$$

(if $h = 0$, then $\alpha_h = 0$, and $\xi = 0$).

If $h \neq 0$, then

$$\xi = \gamma_n \left(\frac{{}^{k_h}D_{*-1}^{\alpha_h} x^h}{h!} - 1 \right) = \gamma_n \left(\int_{-1}^{x} k_h(x-t)\, dt - 1 \right) = \tag{3.61}$$

$$\gamma_n \left(\int_{0}^{x+1} k_h(z)\, dz - 1 \right) \geq \gamma_n \left(\int_{0}^{1} k_h(z)\, dz - 1 \right) \geq 0,$$

by the assumption (3.21): $\int_0^1 k_h(z)\, dz \geq 1$, when $h \neq 0$.

Hence in both cases we get

$$L^*(Q_n(x)) \geq 0, \ x \in [0, 1]. \tag{3.62}$$

II. Suppose on $[0, 1]$ that $\alpha_h(x) \leq \beta < 0$.

Let $Q_n(x) \in S_m(\Delta_n([-1, 1]))$, $x \in [-1, 1]$, as in Lemma 3.5, such that

$$\max_{-1 \leq x \leq 1} \left| {}^{k_j}D_{*-1}^{\alpha_j} \left(f(x) - \gamma_n (h!)^{-1} x^h \right) - \left({}^{k_j}D_{*-1}^{\alpha_j} Q_n \right)(x) \right| \overset{(3.52.)}{\leq} \tag{3.63}$$

$$\lambda_j \frac{C_2^*}{n^{\sigma-j-1}} \omega_{m-\sigma+1}\left(f^{(\sigma-1)}, \frac{1}{n}\right), \quad j = 0, 1, ..., \sigma - 2, \ n \geq 1.$$

Similarly, we obtain again the inequalities of convergence, see (3.46), (3.47) and (3.48).

Also if $0 \leq x \leq 1$, then

$$\alpha_h^{-1}(x) L^*(Q_n(x)) = \alpha_h^{-1}(x) L^*(f(x)) - \frac{\gamma_n}{h!} {}^{k_h}D_{*-1}^{\alpha_h} x^h +$$

$$\sum_{j=h}^{v} \alpha_h^{-1}(x) \alpha_j(x) \left[{}^{k_j}D_{*-1}^{\alpha_j} Q_n(x) - {}^{k_j}D_{*-1}^{\alpha_j} f(x) + \frac{\gamma_n}{h!} {}^{k_j}D_{*-1}^{\alpha_j} x^h \right] \overset{(3.63.)}{\leq}$$

(by $L^* f \geq 0$)

$$-\frac{\gamma_n}{h!} {}^{k_h}D_{*-1}^{\alpha_h} x^h + C_2^* \omega_{m-\sigma+1}\left(f^{(\sigma-1)}, \frac{1}{n}\right) \left(\sum_{j=h}^{v} s_j \lambda_j n^{j-\sigma+1} \right) =$$

$$-\frac{\gamma_n}{h!} {}^{k_h}D_{*-1}^{\alpha_h} x^h + \gamma_n =: \theta \tag{3.64}$$

(if $h = 0$, then $\alpha_h = 0$, and $\theta = 0$).

If $h \neq 0$, then

$$\theta = \gamma_n \left[1 - \frac{{}^{k_h}D_{*-1}^{\alpha_h} x^h}{h!} \right] = \gamma_n \left[1 - \int_{-1}^{x} k_h(x-t)\, dt \right] =$$

$$\gamma_n \left[1 - \int_{0}^{x+1} k_h(z)\, dz \right] \leq \gamma_n \left[1 - \int_{0}^{1} k_h(z)\, dz \right] \leq 0.$$

Hence again in both cases it holds

$$L^*(Q_n(x)) \geq 0, \ \forall\, x \in [0, 1]. \tag{3.65}$$

□ ■

We also give

Theorem 3.13 *Let h, v, σ, m be integers, $2 \leq \sigma \leq m$, $n \in \mathbb{N}$, with h even, $0 \leq h \leq v \leq \sigma - 2$ and let $f \in C^{\sigma-1}([-1, 1])$, with $f^{(\sigma-1)}$ having modulus of smoothness $\omega_{m-\sigma+1}\left(f^{(\sigma-1)}, \delta\right)$ there, $\delta > 0$. Let $\alpha_j(x)$, $j = h, h+1, ..., v$ be real functions, defined and bounded on $[-1, 1]$ and suppose $\alpha_h(x)$ is either $\geq \alpha > 0$ or $\leq \beta < 0$ on $[-1, 0]$. Let the real numbers $\alpha_0 = 0 < \alpha_1 < 1 < \alpha_2 < 2 < ... < \alpha_{\sigma-2} < \sigma - 2$.*

Here ${}^{k_j}D_{1-}^{\alpha_j} f$ *stands for the abstract right Caputo fractional derivative of* f *of order* α_j *anchored at* 1. *We consider the abstract right fractional linear differential operator*

$$L^* := \sum_{j=h}^{v} \alpha_j(x) \left[{}^{k_j}D_{1-}^{\alpha_j} \right], \tag{3.66}$$

and suppose, throughout $[-1, 0]$,

$$L^*(f) \geq 0. \tag{3.67}$$

Set

$$\lambda_j := \int_0^2 k_j(z)\, dz, \ j = 1, ..., \sigma - 2, \tag{3.68}$$

and $\lambda_0 = 1$; *along with*

$$s_j := \sup_{-1 \leq x \leq 1} \left| \alpha_h^{-1}(x)\, \alpha_j(x) \right|, \ j = h, ..., v. \tag{3.69}$$

Then, for any $n \in \mathbb{N}$ *there exists a spline* $Q_n(x) \in S_m(\Delta_n[-1, 1])$ *such that*

$$L^*(Q_n) \geq 0, \ \text{throughout} \ [-1, 0], \tag{3.70}$$

and

(i)

$$\max_{-1 \leq x \leq 1} \left| \left({}^{k_j}D_{1-}^{\alpha_j} f \right)(x) - \left({}^{k_j}D_{1-}^{\alpha_j} Q_n \right)(x) \right| \leq \lambda_j \frac{C_2^*}{n^{\sigma - j - 1}} \omega_{m-\sigma+1}\left(f^{(\sigma-1)}, \frac{1}{n} \right), \tag{3.71}$$

when $j = h + 1, ..., \sigma - 2$;

(ii) *if* $h \neq 0$ *and* $j = 1, ..., h$, *we have*

$$\max_{-1 \leq x \leq 1} \left| \left({}^{k_j}D_{1-}^{\alpha_j} f \right)(x) - \left({}^{k_j}D_{1-}^{\alpha_j} Q_n \right)(x) \right| \leq$$

$$\frac{C_2^* \omega_{m-\sigma+1}\left(f^{(\sigma-1)}, \frac{1}{n} \right)}{n^{\sigma-1-v}} \left[\frac{\left(\sum_{j_*=h}^{v} s_{j_*} \lambda_{j_*} \right)}{h!} \max_{-1 \leq x \leq 1} \left| \left({}^{k_j}D_{1-}^{\alpha_j} x^h \right) \right| + \lambda_j \right]; \tag{3.72}$$

(iii) *furthermore*

$$\max_{-1\leq x\leq 1}|f(x)-Q_n(x)|\leq \frac{C_2^*\omega_{m-\sigma+1}\left(f^{(\sigma-1)},\frac{1}{n}\right)}{n^{\sigma-1-v}}\left[\frac{\left(\sum_{j=h}^{v}s_j\lambda_j\right)}{h!}+1\right]. \quad (3.73)$$

Proof Let Q_n be as in Lemma 3.5. Let also $\alpha_j>0$, $j=1,...,\sigma-2$, such that $\alpha_0=0<\alpha_1<1<\alpha_2<2<...<\alpha_{\sigma-2}<\sigma-2$. That is $\lceil\alpha_j\rceil=j$, $j=0,1,...,\sigma-2$.

We consider the abstract right Caputo fractional derivatives ${}^{k_j}D_{1-}^{\alpha_j}f$ and ${}^{k_j}D_{1-}^{\alpha_j}Q_n$, $j=1,...,\sigma-2$.

We notice that ($x\in[-1,1]$)

$$\left|\left({}^{k_j}D_{1-}^{\alpha_j}f\right)(x)-\left({}^{k_j}D_{1-}^{\alpha_j}Q_n\right)(x)\right|=$$

$$\left|(-1)^j\int_x^1 k_j(t-x)f^{(j)}(t)\,dt-(-1)^j\int_x^1 k_j(t-x)Q_n^{(j)}(t)\,dt\right|= \quad (3.74)$$

$$\left|\int_x^1 k_j(t-x)\left(f^{(j)}(t)-Q_n^{(j)}(t)\right)dt\right|\leq$$

$$\int_x^1 k_j(t-x)\left|f^{(j)}(t)-Q_n^{(j)}(t)\right|dt\overset{(3.9)}{\leq}$$

$$\left(\int_x^1 k_j(t-x)\right)\frac{C_2^*}{n^{\sigma-j-1}}\omega_{m-\sigma+1}\left(f^{(\sigma-1)},\frac{1}{n}\right)=$$

$$\left(\int_0^{1-x} k_j(z)\,dz\right)\frac{C_2^*}{n^{\sigma-j-1}}\omega_{m-\sigma+1}\left(f^{(\sigma-1)},\frac{1}{n}\right)\leq$$

$$\left(\int_0^{2} k_j(z)\,dz\right)\frac{C_2^*}{n^{\sigma-j-1}}\omega_{m-\sigma+1}\left(f^{(\sigma-1)},\frac{1}{n}\right). \quad (3.75)$$

We have proved that

$$\left|\left({}^{k_j}D_{1-}^{\alpha_j}f\right)(x)-\left({}^{k_j}D_{1-}^{\alpha_j}Q_n\right)(x)\right|\leq$$

$$\left(\int_0^2 k_j(z)\,dz\right)\frac{C_2^*}{n^{\sigma-j-1}}\omega_{m-\sigma+1}\left(f^{(\sigma-1)},\frac{1}{n}\right), \tag{3.76}$$

$\forall\, x\in[-1,1]$, $j=1,...,\sigma-2$, $n\geq 1$.

Hence it holds

$$\max_{-1\leq x\leq 1}\left|\left({}^{k_j}D_{1-}^{\alpha_j}f\right)(x)-\left({}^{k_j}D_{1-}^{\alpha_j}Q_n\right)(x)\right|\leq \tag{3.77}$$

$$\lambda_j\frac{C_2^*}{n^{\sigma-j-1}}\omega_{m-\sigma+1}\left(f^{(\sigma-1)},\frac{1}{n}\right),\ j=1,...,\sigma-2,\ n\geq 1,$$

where

$$\lambda_j:=\int_0^2 k_j(z)\,dz,\ j=1,...,\sigma-2,\ \text{as in (3.68.)}.$$

Inequality (3.77) is valid when $j=0$ by (3.9) and (3.20), and we can set $\lambda_0=1$. Set

$$\gamma_n:=C_2^*\omega_{m-\sigma+1}\left(f^{(\sigma-1)},\frac{1}{n}\right)\left(\sum_{j=h}^{v}s_j\lambda_j n^{j-\sigma+1}\right). \tag{3.78}$$

I. Suppose, throughout $[-1,0]$, $\alpha_h(x)\geq\alpha>0$.

Let $Q_n(x)\in S_m(\Delta_n([-1,1]))$, $x\in[-1,1]$, as in Lemma 3.5, so that

$$\max_{-1\leq x\leq 1}\left|{}^{k_j}D_{1-}^{\alpha_j}\left(f(x)+\gamma_n(h!)^{-1}x^h\right)-\left({}^{k_j}D_{1-}^{\alpha_j}Q_n\right)(x)\right|\overset{(3.77.)}{\leq} \tag{3.79}$$

$$\lambda_j\frac{C_2^*}{n^{\sigma-j-1}}\omega_{m-\sigma+1}\left(f^{(\sigma-1)},\frac{1}{n}\right),\ j=0,1,...,\sigma-2.$$

When $j=h+1,...,\sigma-2$, we get

$$\max_{-1\leq x\leq 1}\left|\left({}^{k_j}D_{1-}^{\alpha_j}f\right)(x)-\left({}^{k_j}D_{1-}^{\alpha_j}Q_n\right)(x)\right|\leq\lambda_j\frac{C_2^*}{n^{\sigma-j-1}}\omega_{m-\sigma+1}\left(f^{(\sigma-1)},\frac{1}{n}\right), \tag{3.80}$$

proving (3.71).

For $j=1,...,h\neq 0$, we have that

$$\max_{-1\leq x\leq 1}\left|\left({}^{k_j}D_{1-}^{\alpha_j}f\right)(x)-\left({}^{k_j}D_{1-}^{\alpha_j}Q_n\right)(x)\right|\overset{(3.79.)}{\leq}$$

$$\frac{\gamma_n}{h!} \max_{-1 \le x \le 1} \left| \left({}^{k_j} D_{1-}^{\alpha_j} x^h \right) \right| + \lambda_j \frac{C_2^*}{n^{\sigma-j-1}} \omega_{m-\sigma+1} \left(f^{(\sigma-1)}, \frac{1}{n} \right) =$$

$$\frac{C_2^*}{h!} \omega_{m-\sigma+1} \left(f^{(\sigma-1)}, \frac{1}{n} \right) \left(\sum_{j_*=h}^{v} s_{j_*} \lambda_{j_*} n^{j_*-\sigma+1} \right) \max_{-1 \le x \le 1} \left| \left({}^{k_j} D_{1-}^{\alpha_j} x^h \right) \right| +$$

$$\lambda_j \frac{C_2^*}{n^{\sigma-j-1}} \omega_{m-\sigma+1} \left(f^{(\sigma-1)}, \frac{1}{n} \right) = \tag{3.81}$$

$$C_2^* \omega_{m-\sigma+1} \left(f^{(\sigma-1)}, \frac{1}{n} \right) \left[\frac{1}{h!} \left(\sum_{j_*=h}^{v} s_{j_*} \lambda_{j_*} n^{j_*-\sigma+1} \right) \max_{-1 \le x \le 1} \left| \left({}^{k_j} D_{1-}^{\alpha_j} x^h \right) \right| + \frac{\lambda_j}{n^{\sigma-j-1}} \right] \le$$

$$\frac{C_2^* \omega_{m-\sigma+1} \left(f^{(\sigma-1)}, \frac{1}{n} \right)}{n^{\sigma-1-v}} \left[\frac{\left(\sum_{j_*=h}^{v} s_{j_*} \lambda_{j_*} \right)}{h!} \max_{-1 \le x \le 1} \left| \left({}^{k_j} D_{1-}^{\alpha_j} x^h \right) \right| + \lambda_j \right],$$

proving (3.72).

When $j = 0$ from (3.79) we get

$$\max_{-1 \le x \le 1} \left| f(x) + \gamma_n (h!)^{-1} x^h - Q_n(x) \right| \le \frac{C_2^*}{n^{\sigma-1}} \omega_{m-\sigma+1} \left(f^{(\sigma-1)}, \frac{1}{n} \right). \tag{3.82}$$

That is

$$\max_{-1 \le x \le 1} \left| f(x) - Q_n(x) \right| \le \frac{\gamma_n}{h!} + \frac{C_2^*}{n^{\sigma-1}} \omega_{m-\sigma+1} \left(f^{(\sigma-1)}, \frac{1}{n} \right) =$$

$$\frac{C_2^* \omega_{m-\sigma+1} \left(f^{(\sigma-1)}, \frac{1}{n} \right)}{h!} \left(\sum_{j=h}^{v} s_j \lambda_j n^{j-\sigma+1} \right) + \frac{C_2^*}{n^{\sigma-1}} \omega_{m-\sigma+1} \left(f^{(\sigma-1)}, \frac{1}{n} \right) \le \tag{3.83}$$

$$C_2^* \omega_{m-\sigma+1} \left(f^{(\sigma-1)}, \frac{1}{n} \right) \left[\frac{1}{h! n^{\sigma-1-v}} \left(\sum_{j=h}^{v} s_j \lambda_j \right) + \frac{1}{n^{\sigma-1}} \right] \le$$

$$\frac{C_2^* \omega_{m-\sigma+1} \left(f^{(\sigma-1)}, \frac{1}{n} \right)}{n^{\sigma-1-v}} \left[\frac{\left(\sum_{j=h}^{v} s_j \lambda_j \right)}{h!} + 1 \right], \tag{3.84}$$

proving (3.73).

Also if $-1 \leq x \leq 0$, then

$$\alpha_h^{-1}(x) L^*(Q_n(x)) = \alpha_h^{-1}(x) L^*(f(x)) + \frac{\gamma_n}{h!} {}^{k_h}D_{1-}^{\alpha_h} x^h + \tag{3.85}$$

$$\sum_{j=h}^{v} \alpha_h^{-1}(x) \alpha_j(x) \left[{}^{k_j}D_{1-}^{\alpha_j} Q_n(x) - {}^{k_j}D_{1-}^{\alpha_j} f(x) - \frac{\gamma_n}{h!} {}^{k_j}D_{1-}^{\alpha_j} x^h \right] \overset{(3.79.)}{\geq}$$

(by $L^* f \geq 0$)

$$\frac{\gamma_n}{h!} {}^{k_h}D_{1-}^{\alpha_h} x^h - C_2^* \omega_{m-\sigma+1}\left(f^{(\sigma-1)}, \frac{1}{n}\right) \left(\sum_{j=h}^{v} s_j \lambda_j n^{j-\sigma+1} \right) =$$

$$\frac{\gamma_n}{h!} {}^{k_h}D_{1-}^{\alpha_h} x^h - \gamma_n =: \xi_1$$

(if $h = 0$, then $\alpha_h = 0$, and $\xi_1 = 0$).

If $h \neq 0$, then

$$\xi_1 = \gamma_n \left(\frac{{}^{k_h}D_{1-}^{\alpha_h} x^h}{h!} - 1 \right) = \gamma_n \left((-1)^h \int_x^1 k_h(t-x)\, dt - 1 \right) \overset{(h \text{ is even})}{=} \tag{3.86}$$

$$\gamma_n \left(\int_0^{1-x} k_h(z)\, dz - 1 \right) \geq \gamma_n \left(\int_0^1 k_h(z)\, dz - 1 \right) \geq 0,$$

by the assumption (3.21): $\int_0^1 k_h(z)\, dz \geq 1$, when $h \neq 0$.

Hence in both cases we get

$$L^*(Q_n(x)) \geq 0, \ x \in [-1, 0]. \tag{3.87}$$

II. Suppose on $[-1, 0]$ that $\alpha_h(x) \leq \beta < 0$.

Let $Q_n(x) \in S_m(\Delta_n([-1, 1]))$, $x \in [-1, 1]$, as in Lemma 3.5, so that

$$\max_{-1 \leq x \leq 1} \left| {}^{k_j}D_{1-}^{\alpha_j} \left(f(x) - \gamma_n (h!)^{-1} x^h \right) - \left({}^{k_j}D_{1-}^{\alpha_j} Q_n \right)(x) \right| \overset{(3.77.)}{\leq} \tag{3.88}$$

$$\lambda_j \frac{C_2^*}{n^{\sigma-j-1}} \omega_{m-\sigma+1}\left(f^{(\sigma-1)}, \frac{1}{n}\right), \ j = 0, 1, ..., \sigma - 2, \ n \geq 1.$$

Similarly, we obtain again the inequalities of convergence, see (3.71), (3.72) and (3.73).

Also if $-1 \leq x \leq 0$, then

$$\alpha_h^{-1}(x) L^*(Q_n(x)) = \alpha_h^{-1}(x) L^*(f(x)) - \frac{\gamma_n}{h!} {}^{k_h}D_{1-}^{\alpha_h} x^h +$$

$$\sum_{j=h}^{v} \alpha_h^{-1}(x) \alpha_j(x) \left[{}^{k_j}D_{1-}^{\alpha_j} Q_n(x) - {}^{k_j}D_{1-}^{\alpha_j} f(x) + \frac{\gamma_n}{h!} {}^{k_j}D_{1-}^{\alpha_j} x^h \right] \overset{(3.88.)}{\leq}$$

(by $L^* f \geq 0$)

$$-\frac{\gamma_n}{h!} {}^{k_h}D_{1-}^{\alpha_h} x^h + C_2^* \omega_{m-\sigma+1}\left(f^{(\sigma-1)}, \frac{1}{n}\right)\left(\sum_{j=h}^{v} s_j \lambda_j n^{j-\sigma+1}\right) =$$

$$-\frac{\gamma_n}{h!} {}^{k_h}D_{1-}^{\alpha_h} x^h + \gamma_n =: \theta_1 \tag{3.89}$$

(if $h = 0$, then $\alpha_h = 0$, and $\theta_1 = 0$).

If $h \neq 0$, then

$$\theta_1 = \gamma_n \left[1 - \frac{{}^{k_h}D_{1-}^{\alpha_h} x^h}{h!}\right] = \gamma_n \left[1 - \int_x^1 k_h(t-x)\,dt\right] =$$

$$\gamma_n \left[1 - \int_0^{1-x} k_h(z)\,dz\right] \leq \gamma_n \left[1 - \int_0^1 k_h(z)\,dz\right] \leq 0.$$

Hence again in both cases it holds

$$L^*(Q_n(x)) \geq 0, \ \forall\, x \in [-1, 0]. \tag{3.90}$$

Comment 3.14 *Clearly Theorem 3.12 generalizes Theorem 3.7, and Theorem 3.13 generalizes Theorem 6.2, p. 57 of [2]. Furthermore there, the approximating spline Q_n depends on f, γ_n, h; which γ_n depends on $n, C_2^*, v, s_j, \lambda_j, \sigma$; which λ_j depends on k_j. I.e. polynomial spline Q_n among others depends on the type of fractional calculus we use.*

Consequently, Theorem 3.12 is valid for the following left fractional linear differential operators:

(1)

$$L_1^* := \sum_{j=h}^{v} \alpha_j(x) \left[{}^{C}D_{\rho,\alpha_j,\omega,-1+}^{\gamma} \right], \tag{3.91}$$

where $\rho > 0$, $\gamma < 0$, *and* $\omega > 0$ *large enough (from Prabhakar fractional calculus, see (3.26));*

(2)

$$L_2^* := \sum_{j=h}^{v} \alpha_j(x)\left[D_{-1*}^{\alpha_j}\right], \tag{3.92}$$

(see (3.33)) where $\gamma_j > 0$, $j = 1, ..., m$; $\lambda = 1$; *and small enough* $\omega_j < 0$, $j = 1, ..., m$ *(from generalized non-singular fractional calculus);*

and

(3)

$$L_3^* := \sum_{j=h}^{v} \alpha_j(x)\left[{}_{\omega}^{CF}D_{-1+}^{\alpha_j}\right], \tag{3.93}$$

with $\omega < 0$, *sufficiently small (from parametrized Caputo-Fabrizio non-singular kernel fractional calculus).*

Similarly, Theorem 3.13 is valid for the following right fractional linear differential operators:

(1) *

$$L_1^{**} := \sum_{j=h}^{v} \alpha_j(x)\left[{}^{C}D_{\rho,\alpha_j,\omega,1-}^{\gamma}\right], \tag{3.94}$$

where $\rho > 0$, $\gamma < 0$, *and* $\omega > 0$ *large enough (from Prabhakar fractional calculus, see (3.27));*

(2) *

$$L_2^{**} := \sum_{j=h}^{v} \alpha_j(x)\left[D_{1-}^{\alpha_j}\right], \tag{3.95}$$

(see (3.34)) where $\gamma_j > 0$, $j = 1, ..., m$; $\lambda = 1$; *and small enough* $\omega_j < 0$, $j = 1, ..., m$ *(from generalized non-singular fractional calculus);*

and

(3) *

$$L_3^{**} := \sum_{j=h}^{v} \alpha_j(x)\left[{}_{\omega}^{CF}D_{1-}^{\alpha_j}\right], \tag{3.96}$$

with $\omega < 0$, *sufficiently small (from parametrized Caputo-Fabrizio non-singular kernel fractional calculus).*

Our developed abstract simultaneous fractional spline monotone approximation theory with its applications involves weaker conditions than the one with ordinary derivatives ([1]), and can cover many diverse general cases in a multitude of complex settings and environments.

References

1. Anastassiou, G.A.: Spline monotone approximation with linear differential operators. Approx. Theory Appl. **5**(4), 61–67 (1989)
2. Anastassiou, G.A.: Frontiers in Approximation Theory. World Scientific Publ. Corp, New Jersey (2015)
3. Anastassiou, G.A.: Foundations of Generalized Prabhakar-Hilfer fractional Calculus with Applications. Cubo, accepted (2021)
4. Anastassiou, G.A.: Multiparameter Fractional Differentiation with Non Singular Kernel. Issues of Analysis, accepted (2021)
5. Anastassiou, G.A.: Simultaneous High Order Abstract Fractional Polynomial Spline Monotone Approximation and Applications. Submitted (2021)
6. Atangana, A., Baleanu, D.: New fractional derivatives with nonlocal and non-singular kernel: theory and application to heat transfer model. Therm. Sci. **20**(2), 763–769 (2016)
7. De Vore, R., Lorentz, G.: Constructive Approximation. Springer, Heidelberg, New York (1993)
8. Diethelm, K.: The Analysis of Fractional Differential Equations. Lecture Notes in Mathematics, Vol. 2004, 1st edn. Springer, New York, Heidelberg (2010)
9. Giusti, A., et al.: A practical Guide to Prabhakar Fractional Calculus. Fract. Calculus Appl. Anal. **23**(1), 9–54 (2020)
10. Gorenflo, R., Kilbas, A., Mainardi, F., Rogosin, S.: Mittag-Leffler functions, Related Topics and Applications. Springer, Heidelberg (2014)
11. Hewitt, E., Stromberg, K.: Real and Abstract Analysis. Springer, New York (1965)
12. Losada, J., Nieto, J.J.: Properties of a New Fractional Derivative without Singular Kernel. Progr. Fract. Differ. Appl. **1**(2), 87–92 (2015)
13. Polito, F., Tomovski, Z.: Some properties of Prabhakar-type fractional calculus operators. Fractional Differ. Calculus **6**(1), 73–94 (2016)
14. Saxena, R.K., Kalla, S.L.: Ravi Saxena. Multivariate analogue of generalized Mittag-Leffler function. Integr. Transforms Special Funct. **22**(7), 533–548 (2011)
15. Schumaker, I.L.: Spline Functions: Basic Theory. Wiley, New York (1981)
16. Srivastava, H.M., Daoust, M.C.: A note on the convergence of Kompe' de Feriet's double hypergeometric series. Math. Nachr. **53**, 151–159 (1972)

Chapter 4
Abstract Bivariate Left Fractional Monotone Constrained Approximation by Pseudo-polynomials

Here we extend our earlier bivariate high order simultaneous fractional monotone constrained approximation theory by pseudo-polynomials to abstract bivariate high order simultaneous fractional monotone constrained approximation by pseudo-polynomials, with applications to bivariate Prabhakar fractional calculus and non-singular kernel fractional calculi. We cover the left side of this constrained approximation.

So we deal with the following general two-dimensional problem: Let f be a two variable continuously differentiable real valued function of a given order, let L^* be a linear left abstract fractional mixed partial differential operator and suppose that $L^*(f) \geq 0$ on a critical region. Then for specific and sufficiently large $n, m \in \mathbb{N}$, we can find a sequence of pseudo-polynomials $Q^*_{n,m}$ in two variables with the property $L^*\left(Q^*_{n,m}\right) \geq 0$ on this critical region such that f is approximated with rates fractionally and simultaneously by $Q^*_{n,m}$ in the uniform norm on the whole domain of f. This constrained approximation is given via inequalities involving the mixed modulus of smoothness $\omega_{s,q}$, $s, q \in \mathbb{N}$, of highest order integer partial derivative of f. It follows [7].

4.1 Introduction

The topic of monotone approximation theory started in [21] and it has become a major trend of approximation theory. A typical problem in this subject is: given a positive integer k, approximate a given function whose kth derivative is ≥ 0 by polynomials having this property.

In [8] the authors replaced the kth derivative with a linear differential operator of order k. We mention this motivating result.

G. A. Anastassiou, *Abstract Fractional Monotone Approximation, Theory and Applications*, Studies in Systems, Decision and Control 411,
https://doi.org/10.1007/978-3-030-95943-2_4

Theorem 4.1 *Let h, k, p be integers, $0 \leq h \leq k \leq p$ and let f be a real function, $f^{(p)}$ continuous in $[-1, 1]$ with modulus of continuity $\omega_1\left(f^{(p)}, x\right)$ there. Let $a_j(x)$, $j = h, h+1, ..., k$ be real functions, defined and bounded on $[-1, 1]$ and assume $a_h(x)$ is either $\geq$ some number $\alpha > 0$ or $\leq$ some number $\beta < 0$ throughout $[-1, 1]$. Consider the operator*

$$L = \sum_{j=h}^{k} a_j(x)\left[\frac{d^j}{dx^j}\right] \tag{4.1}$$

and suppose, throughout $[-1, 1]$,

$$L(f) \geq 0. \tag{4.2}$$

Then, for every integer $n \geq 1$, there is a real polynomial $Q_n(x)$ of degree $\leq n$ such that

$$L(Q_n) \geq 0 \text{ throughout } [-1, 1] \tag{4.3}$$

and

$$\max_{-1 \leq x \leq 1} |f(x) - Q_n(x)| \leq C n^{k-p} \omega_1\left(f^{(p)}, \frac{1}{n}\right), \tag{4.4}$$

where C is independent of n or f.

Next let $n, m \in \mathbb{Z}_+$, P_θ denote the space of algebraic polynomials of degree $\leq \theta$. Consider the tensor product spaces $P_n \otimes C([-1, 1])$, $C([-1, 1]) \otimes P_m$ and their sum $P_n \otimes C([-1, 1]) + C([-1, 1]) \otimes P_m$, that is

$$P_n \otimes C([-1, 1]) + C([-1, 1]) \otimes P_m =$$

$$\left\{\sum_{i=0}^{n} x^i A_i(y) + \sum_{j=0}^{m} B_j(x) y^j;\ A_i, B_j \in C([-1, 1]),\ x, y \in [-1, 1]\right\}. \tag{4.5}$$

This is the space of pseudo-polynomials of degree $\leq (n, m)$, first introduced by A. Marchaud in 1924–1927 (see [16, 17]). Here $f^{(k,l)}$ denotes $\frac{\partial^{k+l} f}{\partial x^k \partial y^l}$, the (k, l)-partial derivative of f.

Here we consider the space $C^{r,p}\left([-1, 1]^2\right) = \{f : [-1, 1]^2 \to \mathbb{R};\ f^{(k,l)}$ is continuous for $0 \leq k \leq r, 0 \leq l \leq p\}$. Let $f \in C\left([-1, 1]^2\right)$; for $\delta_1, \delta_2 \geq 0$, define the mixed modulus of smoothness of order (s, q), $s, q \in \mathbb{N}$ (see [20, pp. 516–517]) by

$$\omega_{s,q}(f; \delta_1, \delta_2) \equiv \sup\left\{\left|{}_x\Delta_{h_1}^s \circ_y \Delta_{h_2}^q f(x, y)\right| : (x, y),\right. \tag{4.6}$$

$$\left.(x + sh_1, y + qh_2) \in [-1, 1]^2, |h_i| \leq \delta_i, i = 1, 2\right\}.$$

Here

$$ {}_x\Delta^s_{h_1} \circ_y \Delta^q_{h_2} f(x,y) \equiv \sum_{\sigma=0}^{s}\sum_{\mu=0}^{q}(-1)^{s+q-\sigma-\mu}\binom{s}{\sigma}\binom{q}{\mu} f(x+\sigma h_1, y+\mu h_2) \tag{4.7} $$

is a mixed difference of order (s,q).

We mention

Theorem 4.2 (Gonska [11]). *Let $r,p \in \mathbb{Z}_+$, $s,q \in \mathbb{N}$, and $f \in C^{r,p}\left([-1,1]^2\right)$. Let $n,m \in \mathbb{N}$ with $n \geq \max\{4(r+1), r+s\}$ and $m \geq \max\{4(p+1), p+q\}$. Then there exists a linear operator $Q_{n,m}$ from $C^{r,p}\left([-1,1]^2\right)$ into the space of pseudopolynomials $(P_n \otimes C([-1,1]) + C([-1,1]) \otimes P_m)$ such that*

$$ \left|\left(f - Q_{n,m}(f)\right)^{(k,l)}(x,y)\right| \leq \tag{4.8} $$

$$ M_{r,s} \cdot M_{p,q}\left(\Delta_n(x)\right)^{r-k} \cdot \left(\Delta_m(y)\right)^{p-l} \cdot \omega_{s,q}\left(f^{(r,p)}; \Delta_n(x), \Delta_m(y)\right), $$

for all $(0,0) \leq (k,l) \leq (r,p)$, $x,y \in [-1,1]$, where

$$ \Delta_\theta(z) = \frac{\sqrt{1-z^2}}{\theta} + \frac{1}{\theta^2}, \quad \theta = n,m;\ z = x,y \in [-1,1]. \tag{4.9} $$

The constants $M_{r,s}$, $M_{p,q}$, are independent of f, (x,y) and (n,m); they depend only on (r,s), (p,q), respectively.

See also [12], saying that $Q^{(r,p)}_{n,m}(f)$ is continuous on $[-1,1]^2$.

We need the following result which is an easy consequence of the last theorem (see [20, p. 517]).

Corollary 4.3 *Let $r,p \in \mathbb{Z}_+$, $s,q \in \mathbb{N}$, and $f \in C^{r,p}\left([-1,1]^2\right)$. Let $n,m \in \mathbb{N}$ with $n \geq \max\{4(r+1), r+s\}$ and $m \geq \max\{4(p+1), p+q\}$. Then there exists a pseudopolynomial $Q_{n,m} \equiv Q_{n,m}(f) \in (P_n \otimes C([-1,1]) + C([-1,1]) \otimes P_m)$ such that*

$$ \left\|f^{(k,l)} - Q^{(k,l)}_{n,m}\right\|_\infty \leq \frac{\widetilde{C}}{n^{r-k}m^{p-l}} \cdot \omega_{s,q}\left(f^{(r,p)}; \frac{1}{n}, \frac{1}{m}\right), \tag{4.10} $$

for all $(0,0) \leq (k,l) \leq (r,p)$. Here the constant $\widetilde{C}$ depends only on r,p,s,q.

Corollary 4.3 was used in the proof of the motivational result that follows.

Theorem 4.4 ([1]) *Let h_1, h_2, v_1, v_2, r, p be integers, $0 \leq h_1 \leq v_1 \leq r$, $0 \leq h_2 \leq v_2 \leq p$ and let $f \in C^{r,p}\left([-1,1]^2\right)$, with $f^{(r,p)}$ having a mixed modulus of smoothness $\omega_{s,q}\left(f^{(r,p)}; x, y\right)$ there, $s,q \in \mathbb{N}$. Let $\alpha_{i,j}(x,y)$, $i = h_1, h_1+1, \ldots, v_1$; $j = h_2, h_2+1, \ldots, v_2$ be real-valued functions, defined and bounded in $[-1,1]^2$ and suppose $\alpha_{h_1 h_2}$ is either $\geq \alpha > 0$ or $\leq \beta < 0$ throughout $[-1,1]^2$. Take the operator*

$$L = \sum_{i=h_1}^{v_1} \sum_{j=h_2}^{v_2} \alpha_{ij}(x, y) \frac{\partial^{i+j}}{\partial x^i \partial y^j} \tag{4.11}$$

and assume, throughout $[-1, 1]^2$ *that*

$$L(f) \geq 0. \tag{4.12}$$

Then for any integers n, m *with* $n \geq \max\{4(r+1), r+s\}$, $m \geq \max\{4(p+1), p+q\}$, *there exists a pseudopolynomial*

$$Q_{n,m} \in (P_n \otimes C([-1, 1]) + C([-1, 1]) \otimes P_m)$$

such that $L(Q_{m,n}) \geq 0$ *throughout* $[-1, 1]^2$ *and*

$$\left\| f^{(k,l)} - Q_{n,m}^{(k,l)} \right\|_{\infty} \leq \frac{C}{n^{r-v_1} m^{p-v_2}} \cdot \omega_{s,q}\left(f^{(r,p)}; \frac{1}{n}, \frac{1}{m}\right), \tag{4.13}$$

for all $(0, 0) \leq (k, l) \leq (h_1, h_2)$. *Moreover we get*

$$\left\| f^{(k,l)} - Q_{n,m}^{(k,l)} \right\|_{\infty} \leq \frac{C}{n^{r-k} m^{p-l}} \cdot \omega_{s,q}\left(f^{(r,p)}; \frac{1}{n}, \frac{1}{m}\right), \tag{4.14}$$

for all $(h_1+1, h_2+1) \leq (k, l) \leq (r, p)$. *Also (4.14) is valid whenever* $0 \leq k \leq h_1$, $h_2 + 1 \leq l \leq p$ *or* $h_1 + 1 \leq k \leq r$, $0 \leq l \leq h_2$. *Here* C *is a constant independent of* f *and* n, m. *It depends only on* r, p, s, q, L.

We are also motivated by Anastassiou [2].

We mention

Definition 4.5 (*see* [15]) Let $[-1, 1]^2$; $\alpha_1, \alpha_2 > 0$; $\alpha = (\alpha_1, \alpha_2)$, $f \in C([-1, 1]^2)$, $x = (x_1, x_2)$, $t = (t_1, t_2) \in [-1, 1]^2$. We define the left mixed Riemann-Liouville fractional two dimensional integral of order α

$$\left(I_{-1+}^{\alpha} f\right)(x) := \frac{1}{\Gamma(\alpha_1)\Gamma(\alpha_2)} \int_{-1}^{x_1} \int_{-1}^{x_2} (x_1 - t_1)^{\alpha_1 - 1} (x_2 - t_2)^{\alpha_2 - 1} f(t_1, t_2)\, dt_1 dt_2, \tag{4.15}$$

with $x_1, x_2 > -1$.

Notice here that $I_{-1+}^{\alpha}(|f|) < \infty$.

Definition 4.6 ([3, pp. 1–17]) Let $\alpha_1, \alpha_2 > 0$ with $\lceil \alpha_1 \rceil = m_1$, $\lceil \alpha_2 \rceil = m_2$, ($\lceil \cdot \rceil$ ceiling of the number). Let here $f \in C^{m_1, m_2}([-1, 1]^2)$. We consider the left Caputo type fractional partial derivative:

$$D_{*(-1)}^{(\alpha_1, \alpha_2)} f(x) := \frac{1}{\Gamma(m_1 - \alpha_1)\Gamma(m_2 - \alpha_2)}.$$

$$\int_{-1}^{x_1}\int_{-1}^{x_2} (x_1 - t_1)^{m_1-\alpha_1-1} (x_2 - t_2)^{m_2-\alpha_2-1} \frac{\partial^{m_1+m_2} f(t_1, t_2)}{\partial t_1^{m_1} \partial t_2^{m_2}} dt_1 dt_2, \tag{4.16}$$

$\forall\, x = (x_1, x_2) \in [-1, 1]^2$, where Γ is the gamma function

$$\Gamma(\nu) = \int_0^{\infty} e^{-t} t^{\nu-1} dt, \quad \nu > 0. \tag{4.17}$$

We set

$$D_{*(-1)}^{(0,0)} f(x) := f(x), \quad \forall\, x \in [-1, 1]^2; \tag{4.18}$$

$$D_{*(-1)}^{(m_1,m_2)} f(x) := \frac{\partial^{m_1+m_2} f(x)}{\partial x_1^{m_1} \partial x_2^{m_2}}, \quad \forall\, x \in [-1, 1]^2. \tag{4.19}$$

Definition 4.7 We also set

$$D_{*(-1)}^{(0,\alpha_2)} f(x) := \frac{1}{\Gamma(m_2 - \alpha_2)} \int_{-1}^{x_2} (x_2 - t_2)^{m_2-\alpha_2-1} \frac{\partial^{m_2} f(x_1, t_2)}{\partial t_2^{m_2}} dt_2, \tag{4.20}$$

$$D_{*(-1)}^{(\alpha_1,0)} f(x) := \frac{1}{\Gamma(m_1 - \alpha_1)} \int_{-1}^{x_1} (x_1 - t_1)^{m_1-\alpha_1-1} \frac{\partial^{m_1} f(t_1, x_2)}{\partial t_1^{m_1}} dt_1, \tag{4.21}$$

and

$$D_{*(-1)}^{(m_1,\alpha_2)} f(x) := \frac{1}{\Gamma(m_2 - \alpha_2)} \int_{-1}^{x_2} (x_2 - t_2)^{m_2-\alpha_2-1} \frac{\partial^{m_1+m_2} f(x_1, t_2)}{\partial x_1^{m_1} \partial t_2^{m_2}} dt_2, \tag{4.22}$$

$$D_{*(-1)}^{(\alpha_1,m_2)} f(x) := \frac{1}{\Gamma(m_1 - \alpha_1)} \int_{-1}^{x_1} (x_1 - t_1)^{m_1-\alpha_1-1} \frac{\partial^{m_1+m_2} f(t_1, x_2)}{\partial t_1^{m_1} \partial x_2^{m_2}} dt_1. \tag{4.23}$$

In [3, pp. 1–17], we extended Theorem 4.4 to the fractional level. Indeed there L is replaced by L^*, a linear left Caputo fractional mixed partial differential operator. Now the monotonicity property holds true only on the critical square of $[0, 1]^2$. Simultaneously fractional convergence remains true on all of $[-1, 1]^2$.

So we have proved there

Theorem 4.8 *Let h_1, h_2, v_1, v_2, r, p be integers, $0 \le h_1 \le v_1 \le r$, $0 \le h_2 \le v_2 \le p$ and let $f \in C^{r,p}\left([-1, 1]^2\right)$, with $f^{(r,p)}$ having a mixed modulus of smoothness $\omega_{s,q}\left(f^{(r,p)}; x, y\right)$ there, $s, q \in \mathbb{N}$. Let $\alpha_{ij}(x, y)$, $i = h_1, h_1 + 1, ..., v_1$; $j =$*

$h_2, h_2+1, \ldots, v_2$ *be real valued functions, defined and bounded in* $[-1,1]^2$ *and suppose* $\alpha_{h_1 h_2}$ *is either* $\geq \alpha > 0$ *or* $\leq \beta < 0$ *throughout* $[0,1]^2$. *Here* $n, m \in \mathbb{N}$: $n \geq \max\{4(r+1), r+s\}$, $m \geq \max\{4(p+1), p+q\}$. *Set*

$$l_{ij} := \sup_{(x,y)\in[-1,1]^2} \left|\alpha_{h_1 h_2}^{-1}(x,y)\,\alpha_{ij}(x,y)\right| < \infty, \tag{4.24}$$

for all $h_1 \leq i \leq v_1$, $h_2 \leq j \leq v_2$. *Let* $\alpha_{1i}, \alpha_{2j} \geq 0$ *with* $\lceil \alpha_{1i} \rceil = i$, $\lceil \alpha_{2j} \rceil = j$, $i = 0, 1, \ldots, r$; $j = 0, 1, \ldots, p$, *(*$\lceil \cdot \rceil$ *ceiling of the number),* $\alpha_{10} = 0$, $\alpha_{20} = 0$.

Consider the left fractional bivariate differential operator

$$L^* := \sum_{i=h_1}^{v_1} \sum_{j=h_2}^{v_2} \alpha_{ij}(x,y)\, D_{*(-1)}^{(\alpha_{1i},\alpha_{2j})}. \tag{4.25}$$

Assume $L^* f(x,y) \geq 0$*, on* $[0,1]^2$.

Then there exists

$$Q_{n,m}^* \equiv Q_{n,m}^*(f) \in (P_n \otimes C([-1,1]) + C([-1,1]) \otimes P_m)$$

such that $L^* Q_{n,m}^*(x,y) \geq 0$*, on* $[0,1]^2$.

Furthermore it holds:

(1)

$$\left\| D_{*(-1)}^{(\alpha_{1i},\alpha_{2j})}(f) - D_{*(-1)}^{(\alpha_{1i},\alpha_{2j})} Q_{n,m}^* \right\|_{\infty,[-1,1]^2} \leq$$

$$\frac{\widetilde{C}\, 2^{(i+j)-(\alpha_{1i}+\alpha_{2j})}}{\Gamma(i-\alpha_{1i}+1)\,\Gamma(j-\alpha_{2j}+1)\, n^{r-i} m^{p-j}} \cdot \omega_{s,q}\left(f^{(r,p)}; \frac{1}{n}, \frac{1}{m}\right), \tag{4.26}$$

where $\widetilde{C}$ *is a constant that depends only on* r, p, s, q; $(h_1+1, h_2+1) \leq (i,j) \leq (r,p)$*, or* $0 \leq i \leq h_1$, $h_2+1 \leq j \leq p$*, or* $h_1+1 \leq i \leq r$, $0 \leq j \leq h_2$,

(2)

$$\left\| D_{*(-1)}^{(\alpha_{1i},\alpha_{2j})}(f) - D_{*(-1)}^{(\alpha_{11},\alpha_{2j})} Q_{n,m}^* \right\|_{\infty,[-1,1]^2} \leq$$

$$\frac{c_{ij}}{n^{r-v_1} m^{p-v_2}} \cdot \omega_{s,q}\left(f^{(r,p)}; \frac{1}{n}, \frac{1}{m}\right), \tag{4.27}$$

for $(1,1) \leq (i,j) \leq (h_1,h_2)$, *where* $c_{ij} = \widetilde{C} A_{ij}$, *with*

$$A_{ij} :=$$

$$\left\{ \left[\sum_{\tau=h_1}^{v_1} \sum_{\mu=h_2}^{v_2} \frac{l_{\tau\mu} 2^{(\tau+\mu)-(\alpha_{1\tau}+\alpha_{2\mu})}}{\Gamma(\tau - a_{1\tau}+1)\Gamma(\mu - \alpha_{2\mu}+1)} \right] \left(\sum_{k=0}^{h_1-i} \frac{2^{h_1-\alpha_{1i}-k}}{k!\Gamma(h_1-\alpha_{1i}-k+1)} \right) \cdot \right. \tag{4.28}$$

$$\left. \left(\sum_{\lambda=0}^{h_2-j} \frac{2^{h_2-\alpha_{2j}-\lambda}}{\lambda!\Gamma(h_2-\alpha_{2j}-\lambda+1)} \right) + \frac{2^{(i+j)-(\alpha_{1i}+\alpha_{2j})}}{\Gamma(i-\alpha_{1i}+1)\Gamma(j-\alpha_{2j}+1)} \right\},$$

(3)

$$\left\| f - Q_{n,m}^{*} \right\|_{\infty,[-1,1]^2} \leq \frac{c_{00}}{n^{r-v_1} m^{p-v_2}} \cdot \omega_{s,q}\left(f^{(r,p)}; \frac{1}{n}, \frac{1}{m} \right), \tag{4.29}$$

where $c_{00} := \widetilde{C} A_{00}$, *with*

$$A_{00} := \frac{1}{h_1!h_2!} \left(\sum_{\tau=h_1}^{v_1} \sum_{\mu=h_2}^{v_2} l_{\tau\mu} \frac{2^{(\tau+\mu)-(\alpha_{1\tau}+\alpha_{2\mu})}}{\Gamma(\tau - a_{1\tau}+1)\Gamma(\mu-\alpha_{2\mu}+1)} \right) + 1,$$

(4)

$$\left\| D_{*(-1)}^{(0,\alpha_{2j})}(f) - D_{*(-1)}^{(0,\alpha_{2j})} Q_{n,m}^{*} \right\|_{\infty,[-1,1]^2} \leq$$

$$\frac{c_{0j}}{n^{r-v_1} m^{p-v_2}} \cdot \omega_{s,q}\left(f^{(r,p)}; \frac{1}{n}, \frac{1}{m} \right), \tag{4.30}$$

where $c_{0j} = \widetilde{C} A_{0j}$, $j = 1, ..., h_2$, *with*

$$A_{0j} := \left[\frac{1}{h_1!} \left(\sum_{\tau=h_1}^{v_1} \sum_{\mu=h_2}^{v_2} l_{\tau\mu} \frac{2^{(\tau+\mu)-(\alpha_{1\tau}+\alpha_{2\mu})}}{\Gamma(\tau - a_{1\tau}+1)\Gamma(\mu-\alpha_{2\mu}+1)} \right) \right. \tag{4.31}$$

$$\left. \left(\sum_{\lambda=0}^{h_2-j} \frac{2^{h_2-\alpha_{2j}-\lambda}}{\lambda!\Gamma(h_2-\alpha_{2j}-\lambda+1)} \right) + \frac{2^{j-\alpha_{2j}}}{\Gamma(j-\alpha_{2j}+1)} \right],$$

and

(5)

$$\left\| D_{*(-1)}^{(\alpha_{1i},0)}(f) - D_{*(-1)}^{(\alpha_{1i},0)} Q_{n,m}^{*} \right\|_{\infty,[-1,1]^2} \leq$$

$$\frac{c_{i0}}{n^{r-v_1} m^{p-v_2}} \cdot \omega_{s,q}\left(f^{(r,p)}; \frac{1}{n}, \frac{1}{m} \right), \tag{4.32}$$

where $c_{i0} = \widetilde{C} A_{i0}$, $i = 1, ..., h_1$, *with*

$$A_{i0} := \left[\frac{1}{h_2!} \left(\sum_{\tau=h_1}^{v_1} \sum_{\mu=h_2}^{v_2} l_{\tau\mu} \frac{2^{(\tau+\mu)-(\alpha_{1\tau}+\alpha_{2\mu})}}{\Gamma(\tau - a_{1\tau} + 1)\Gamma(\mu - \alpha_{2\mu} + 1)} \right) \cdot \right. \quad (4.33)$$

$$\left. \left(\sum_{k=0}^{h_1 - i} \frac{2^{h_1 - \alpha_{1i} - k}}{k!\Gamma(h_1 - \alpha_{1i} - k + 1)} \right) + \frac{2^{i-\alpha_{1i}}}{\Gamma(i - \alpha_{1i} + 1)} \right].$$

In this chapter we establish bivariate abstract fractional calculus left high order monotone (constrained) approximation theory by pseudo-polynomials of Caputo type, and then we apply our results to bivariate Prabhakar fractional Calculus, bivariate generalized non-singular fractional calculus, and bivariate parametrized Caputo-Fabrizio non-singular fractional calculus.

Next, we build the related necessary fractional calculi background.

4.2 Bivariate Fractional Calculi

Here we need to be very specific in preparation for our main results.

4.2.1 Bivariate Abstract Fractional Calculus

Let h_1, h_2, v_1, v_2, r, p be integers, $0 \leq h_1 \leq v_1 \leq r$, $0 \leq h_2 \leq v_2 \leq p$ and let $f \in C^{r,p}([-1, 1]^2)$. Here $h_1 \leq i \leq v_1$, $h_2 \leq j \leq v_2$. Let $\alpha_{1i}, \alpha_{2j} \geq 0$, $\alpha_{1i}, \alpha_{2j} \notin \mathbb{Z}_+ :$ $\lceil \alpha_{1i} \rceil = i$, $\lceil \alpha_{2j} \rceil = j$, $i = 0, 1, ..., r$; $j = 0, 1, ..., p$, ($\lceil \cdot \rceil$ is the ceiling of number), $\alpha_{10} = 0$, $\alpha_{20} = 0$.

Consider also the integrable functions $k_{1i} := K_{\alpha_{1i}}, k_{2j} := K_{\alpha_{2j}} : [0, 2] \to \mathbb{R}_+$, $i = 0, 1, ..., r$; $j = 0, 1, ..., p$.

We consider the abstract left Caputo type bivariate fractional partial derivative of orders $(\alpha_{1i}, \alpha_{2j})$

$$ {}^{k_{1i}}_{k_{2j}} D^{(\alpha_{1i}, \alpha_{2j})}_{*(-1)} f(x) := \int_{-1}^{x_1} \int_{-1}^{x_2} k_{1i}(x_1 - t_1) k_{2j}(x_2 - t_2) \frac{\partial^{i+j} f(t_1, t_2)}{\partial t_1^i \partial t_2^j} dt_1 dt_2, \quad (4.34)$$

$\forall\, x = (x_1, x_2) \in [-1, 1]^2$.

We set

$$\begin{gathered} {}^{k_{10}}_{k_{20}} D^{(0,0)}_{*(-1)} f(x) := f(x), \\ {}^{k_{1i}}_{k_{2j}} D^{(i,j)}_{*(-1)} f(x) := \frac{\partial^{i+j} f(x_1, x_2)}{\partial x_1^i \partial x_2^j}, \quad \forall\, x = (x_1, x_2) \in [-1, 1]^2 . \end{gathered} \tag{4.35}$$

We also set

$$ {}^{k_{1i}}_{k_{2j}} D^{(i,\alpha_{2j})}_{*(-1)} f(x) := \int_{-1}^{x_2} k_{2j}(x_2 - t_2) \frac{\partial^{i+j} f(x_1, t_2)}{\partial x_1^i \partial t_2^j} dt_2, \tag{4.36}$$

$$ {}^{k_{1i}}_{k_{2j}} D^{(\alpha_{1i},j)}_{*(-1)} f(x) := \int_{-1}^{x_1} k_{1i}(x_1 - t_1) \frac{\partial^{i+j} f(t_1, x_2)}{\partial t_1^i \partial x_2^j} dt_1, \tag{4.37}$$

and in particular we define:

$$ {}^{k_{10}}_{k_{2j}} D^{(0,\alpha_{2j})}_{*(-1)} f(x) := \int_{-1}^{x_2} k_{2j}(x_2 - t_2) \frac{\partial^{j} f(x_1, t_2)}{\partial t_2^j} dt_2, \tag{4.38}$$

$$ {}^{k_{1i}}_{k_{20}} D^{(\alpha_{1i},0)}_{*(-1)} f(x) := \int_{-1}^{x_1} k_{1i}(x_1 - t_1) \frac{\partial^{i} f(t_1, x_2)}{\partial t_1^i} dt_1, \tag{4.39}$$

$\forall\, x = (x_1, x_2) \in [-1, 1]^2$.

We will assume that

$$\int_0^1 k_{ih_i}(z)\, dz \geq 1, \text{ when } h_i \neq 0, \tag{4.40}$$

where $i = 1, 2$.

In [3], we got that $0 < \Gamma(h_1 - \alpha_{1h_1} + 1), \Gamma(h_2 - \alpha_{2h_2} + 1) \leq 1$, where Γ is the gamma function, and there it is

$$k_{ih_i}(z) = \frac{z^{h_i - \alpha_{ih_i} - 1}}{\Gamma(h_i - \alpha_{ih_i})}, \quad i = 1, 2; \ \forall\, z \in [0, 2],$$

and (4.40) is fulfilled.

4.2.2 About Bivariate Fractional Calculus

We consider the Prabhakar function (also known as the three parameter Mittag-Leffler function), (see [13, p. 97]; [10])

$$E_{\alpha,\beta}^{\gamma}(z) = \sum_{k=0}^{\infty} \frac{(\gamma)_k}{k!\Gamma(\alpha k + \beta)} z^k, \tag{4.41}$$

where $\alpha, \beta > 0$, $\gamma \in \mathbb{R}$, $z \in \mathbb{R}$, and $(\gamma)_k = \gamma(\gamma+1)\ldots(\gamma+k-1)$, it is $E_{\alpha,\beta}^{0}(z) = \frac{1}{\Gamma(\beta)}$.

Let $a, b, c, d \in \mathbb{R}$, $a < b$, $c < d$; $f \in C^{N_1,N_2}([a,b] \times [c,d])$, $\rho_i, \mu_i > 0$; γ_i, $\omega_i \in \mathbb{R}$; $N_i = \lceil \mu_i \rceil$, $\mu_i \notin \mathbb{N}$; $i = 1, 2$.

We define the bivariate Prabhakar-Caputo left partial fractional derivative of orders (μ_1, μ_2) as follows ($x = (x_1, x_2) \in [a,b] \times [c,d]$):

$$\left({}^{C}D_{(\rho_1,\rho_2),(\mu_1,\mu_2),(\omega_1,\omega_2),(a+,c+)}^{(\gamma_1,\gamma_2)} f\right)(x) = \int_a^{x_1}\int_c^{x_2} (x_1-t_1)^{N_1-\mu_1-1}(x_2-t_2)^{N_2-\mu_2-1}$$

$$E_{\rho_1,N_1-\mu_1}^{-\gamma_1}\left[\omega_1(x_1-t_1)^{\rho_1}\right] E_{\rho_2,N_2-\mu_2}^{-\gamma_2}\left[\omega_2(x_2-t_2)^{\rho_2}\right] \frac{\partial^{N_1+N_2} f(t_1,t_2)}{\partial t_1^{N_1}\partial t_2^{N_2}} dt_1 dt_2, \tag{4.42}$$

with

$$\left({}^{C}D_{(\rho_1,\rho_2),(0,0),(\omega_1,\omega_2),(a+,c+)}^{(\gamma_1,\gamma_2)} f\right)(x) := f(x);$$

$$\left({}^{C}D_{(\rho_1,\rho_2),(N_1,N_2),(\omega_1,\omega_2),(a+,c+)}^{(\gamma_1,\gamma_2)} f\right)(x) := \frac{\partial^{N_1+N_2} f(x_1,x_2)}{\partial x_1^{N_1}\partial x_2^{N_2}}, \tag{4.43}$$

where $N_1, N_2 \in \mathbb{N}$, etc.

For the related univariate theory see [5, 18].

4.2.3 About Generalized Non-singular Fractional Calculus

Here we use the multivariate analogue of generalized Mittag-Leffler function, see [19], defined for $\lambda, \gamma_j, \rho_j, z_j \in \mathbb{C}$, $Re(\rho_j) > 0$ $(j = 1, \ldots, m)$ in terms of a multiple series of the form:

$$E_{(\rho_j),\lambda}^{(\gamma_j)}(z_1, \ldots, z_m) = E_{(\rho_1,\ldots,\rho_m),\lambda}^{(\gamma_1,\ldots,\gamma_m)}(z_1, \ldots, z_m) =$$

$$\sum_{k_1,...,k_m=0}^{\infty} \frac{(\gamma_1)_{k_1} \dots (\gamma_m)_{k_m}}{\Gamma\left(\lambda + \sum_{j=1}^{m} k_j \rho_j\right)} \frac{z_1^{k_1} \dots z_m^{k_m}}{k_1! \dots k_m!}, \tag{4.44}$$

where $(\gamma_j)_{k_j}$ is the Pochhammer symbol. By [22, p. 157], (4.44) converges for $Re(\rho_j) > 0$, $j = 1, ..., m$.

In particular we will use $\theta = 1, 2$; $E^{(\gamma_{\theta 1},...,\gamma_{\theta m})}_{(\rho_\theta,...,\rho_\theta),\lambda_\theta}\left[\omega_{\theta 1} t_\theta^{\rho_\theta}, ..., \omega_{\theta m} t_\theta^{\rho_\theta}\right]$, denoted by $E^{(\gamma_{\theta j})}_{(\rho_\theta),\lambda_\theta}\left[\omega_{\theta 1} t_\theta^{\rho_\theta}, ..., \omega_{\theta m} t_\theta^{\rho_\theta}\right]$, where $0 < \rho_\theta < 1$, $t_\theta \geq 0$, $\lambda_\theta > 0$, $\gamma_{\theta j} \in \mathbb{R}$ with $(\gamma_{\theta j})_{k_{0j}} := \gamma_{\theta j}(\gamma_{\theta j} + 1) \dots (\gamma_{\theta j} + k_{\theta j} - 1)$, $\omega_{\theta j} \in \mathbb{R} - \{0\}$, for $j = 1, ..., m$.

Let $f \in C^{N_1,N_2}([-1,1]^2)$, $0 < \mu_\theta \notin \mathbb{N}$ and $\lceil \mu_\theta \rceil = N_\theta \in \mathbb{N}$; $\theta = 1, 2$.

We define the bivariate Caputo type generalized left partial fractional derivative with non-singular kernel of order (μ_1, μ_2), as follows:

$$D_{-1*}^{(\mu_1,\mu_2)} f(x) := {}^{(\gamma_{1j})(\omega_{1j})CA}_{(\gamma_{2j})(\omega_{2j})} D_{-1*}^{(\mu_1,\mu_2),(\lambda_1,\lambda_2)} f(x) := \frac{A(1 - n_1 + \mu_1, 1 - N_2 + \mu_2)}{(N_1 - \mu_1)(N_2 - \mu_2)}$$

$$\int_{-1}^{x_1}\int_{-1}^{x_2} \prod_{\theta=1}^{2} E^{(\gamma_{\theta j})}_{(1-N_\theta+\mu_\theta),\lambda_\theta}\left[\frac{-\omega_{\theta 1}(1 - N_\theta + \mu_\theta)}{N_\theta - \mu_\theta}(x_\theta - t_\theta)^{1-N_\theta+\mu_\theta}\right.$$

$$\left., ..., \frac{-\omega_{\theta m}(1 - N_\theta + \mu_\theta)}{N_\theta - \mu_\theta}(x_\theta - t_\theta)^{1-N_\theta+\mu_\theta}\right] \frac{\partial^{N_1+N_2} f(t_1, t_2)}{\partial t_1^{N_1} \partial t_2^{N_2}} dt_1 dt_2, \tag{4.45}$$

$\forall x = (x_1, x_2) \in [-1, 1]^2$, where $A := A(1 - n_1 + \mu_1, 1 - N_2 + \mu_2)$ is a normalizing constant. Without loss of generality we assume that $A > 0$.

We set $D_{-1*}^{(0,0)} f = f$, $D_{-1*}^{(N_1,N_2)} f = \frac{\partial^{N_1+N_2} f}{\partial x_1^{N_1} \partial x_2^{N_2}}$, when $N_1, N_2 \in \mathbb{N}$, etc.

For the univariate theory see the related [4, 6, 9].

4.2.4 *Bivariate Parametrized Caputo-Fabrizio Type Non-singular Kernel Left Partial Fractional Derivative of Orders* (μ_1, μ_2)

Let $f \in C^{N_1,N_2}([-1,1]^2)$, $\mathbb{N} \not\ni \mu_1, \mu_2 > 0$, $\lceil \mu_\theta \rceil = N_\theta \in \mathbb{N}$; $\omega_\theta < 0$; $\theta = 1, 2$.

It is given by

$${}^{CF}_{(\omega_1,\omega_2)} D_{-1+}^{(\mu_1,\mu_2)} f(x) := \frac{1}{(N_1 - \mu_1)(N_2 - \mu_2)}$$

$$\int_{-1}^{x_1}\int_{-1}^{x_2}\prod_{\theta=1}^{2}\left[\exp\left(-\frac{(1-N_\theta+\mu_\theta)\,\omega_\theta}{N_\theta-\mu_\theta}(x_\theta-t_\theta)\right)\right]\frac{\partial^{N_1+N_2} f(t_1,t_2)}{\partial t_1^{N_1}\partial t_2^{N_2}}dt_1dt_2, \quad (4.46)$$

$\forall\, x=(x_1,x_2)\in[-1,1]^2$, and ${}^{CF}_{(\omega_1,\omega_2)}D^{(0,0)}_{-1+}f=f$, ${}^{CF}_{(\omega_1,\omega_2)}D^{(N_1,N_2)}_{-1+}f=\frac{\partial^{N_1+N_2}f}{\partial x_1^{N_1}\partial x_2^{N_2}}$, etc.

For the univariate case see [6, 14].

We make

Remark 4.9 Fractional Calculi 4.2.2–4.2.4 are special cases of the abstract fractional calculus 4.2.1. The abstract important condition (4.40) is fulfilled by: in Sect. 4.2.2 for large enough $\omega_\theta>0, \theta=1,2$; in Sect. 4.2.3 for small enough $\omega_{\theta j}<0, \theta=1,2$; $j=1,\ldots,m$; and in Sect. 4.2.4 for small enough $\omega_\theta<0$, $\theta=1,2$. For details see [6] and Chap. 1.

4.3 Main Results

We present the following bivariate abstract fractional monotone (constrained) approximation result.

Theorem 4.10 *Let h_1,h_2,v_1,v_2,r,p be integers, $0\le h_1\le v_1\le r$, $0\le h_2\le v_2\le p$ and let $f\in C^{r,p}\left([-1,1]^2\right)$, with $f^{(r,p)}$ having a mixed modulus of smoothness $\omega_{s,q}\left(f^{(r,p)};\delta_1,\delta_2\right)$, $\delta_1,\delta_2>0$ there, $s,q\in\mathbb{N}$. Let $\alpha_{ij}(x,y)$, $i=h_1,h_1+1,\ldots,v_1$; $j=h_2,h_2+1,\ldots,v_2$ be real-valued functions, defined and bounded in $[-1,1]^2$ and suppose $\alpha_{h_1h_2}$ is either $\ge\alpha>0$ or $\le\beta<0$ throughout $[0,1]^2$. Here $n,m\in\mathbb{N}:n\ge\max\{4(r+1),r+s\}$, $m\ge\max\{4(p+1),p+q\}$. Set*

$$l_{ij}:=\sup_{(x,y)\in[-1,1]^2}\left|\alpha_{h_1h_2}^{-1}(x,y)\,\alpha_{ij}(x,y)\right|<\infty \quad (4.47)$$

for all $h_1\le i\le v_1$, $h_2\le j\le v_2$. Let $0<\alpha_{1i},\alpha_{2j}\notin\mathbb{N}$ with $\lceil\alpha_{1i}\rceil=i$, $\lceil\alpha_{2j}\rceil=j$, $j=1,\ldots,r$; $j=1,\ldots,p$ ($\lceil\cdot\rceil$ is the ceiling of the number) and $\alpha_{10}=\alpha_{20}=0$.

Consider the abstract left fractional bivariate differential operator

$$L^*:=\sum_{i=h1}^{v_1}\sum_{j=h_2}^{v_2}\alpha_{ij}(x,y)\;{}^{k_{1i}}_{k_{2j}}D^{(\alpha_{1i},\alpha_{2j})}_{*(-1)}. \quad (4.48)$$

Assume $L^*f(x,y)\ge0$, on $[0,1]^2$. There exists a pseudo-polynomial of degree $\le(n,m)$

$$Q^*_{n,m}:=Q^*_{n,m}(f)\in\left(P_n\otimes C([-1,1])+C([-1,1])\otimes P_m\right)$$

such that $L^*Q^*_{n,m}(x,y)\ge0$, on $[0,1]^2$.

We set

$$\lambda_{1i} := \int_0^2 k_{1i}(z)\,dz, \quad \lambda_{2j} := \int_0^2 k_{2j}(z)\,dz, \tag{4.49}$$

for $i = 1, ..., r;\ j = 1, ..., p$.

Set also $\lambda_{10} = \lambda_{20} = 1$.

Furthermore it holds:

(1) if $(h_1 + 1, h_2 + 1) \le (i, j) \le (r, p)$, *or* $0 \le i \le h_1$, $h_2 + 1 \le j \le p$, *or* $h_1 + 1 \le i \le r$, $0 \le j \le h_2$, *then*

$$\left\| {}^{k_{1i}}_{k_{2j}} D^{(\alpha_{1i},\alpha_{2j})}_{*(-1)} f - {}^{k_{1i}}_{k_{2j}} D^{(\alpha_{1i},\alpha_{2j})}_{*(-1)} Q^*_{n,m} \right\|_{\infty,[-1,1]^2} \le$$

$$\lambda_{1i}\lambda_{2j} \frac{\widetilde{C}}{n^{r-i} m^{p-j}} \omega_{s,q}\left(f^{(r,p)}; \frac{1}{n}, \frac{1}{m}\right), \tag{4.50}$$

(2) if $(1, 1) \le (i, j) \le (h_1, h_2)$, *then*

$$\left\| {}^{k_{1i}}_{k_{2j}} D^{(\alpha_{1i},\alpha_{2j})}_{*(-1)} f - {}^{k_{1i}}_{k_{2j}} D^{(\alpha_{1i},\alpha_{2j})}_{*(-1)} Q^*_{n,m} \right\|_{\infty,[-1,1]^2} \le \frac{\widetilde{C}\omega_{s,q}\left(f^{(r,p)}; \frac{1}{n}, \frac{1}{m}\right)}{n^{r-v_1} m^{p-v_2}}$$

$$\left\{ \left(\frac{\sum_{i_*=h_1}^{v_1} \sum_{j_*=h_2}^{v_2} l_{i_* j_*} \lambda_{1i_*} \lambda_{2j_*}}{h_1! h_2!} \right) \left\| {}^{k_{1i}}_{k_{2j}} D^{(\alpha_{1i},\alpha_{2j})}_{*(-1)} \left(x^{h_1} y^{h_2}\right) \right\|_{\infty,[-1,1]^2} + \lambda_{1i}\lambda_{2j} \right\}, \tag{4.51}$$

(3) it holds

$$\left\| f - Q^*_{n,m} \right\|_{\infty,[-1,1]^2} \le \frac{\widetilde{C}\omega_{s,q}\left(f^{(r,p)}; \frac{1}{n}, \frac{1}{m}\right)}{n^{r-v_1} m^{p-v_2}} \left[\frac{\sum_{i_*=h_1}^{v_1} \sum_{j_*=h_2}^{v_2} l_{i_* j_*} \lambda_{1i_*} \lambda_{2j_*}}{h_1! h_2!} + 1 \right], \tag{4.52}$$

(4) when $i = 0$, $j = 1, ..., h_2$, *we get*

$$\left\| {}^{k_{10}}_{k_{2j}} D^{(0,\alpha_{2j})}_{*(-1)} f - {}^{k_{10}}_{k_{2j}} D^{(0,\alpha_{2j})}_{*(-1)} Q^*_{n,m} \right\|_{\infty,[-1,1]^2} \le \frac{\widetilde{C}\omega_{s,q}\left(f^{(r,p)}; \frac{1}{n}, \frac{1}{m}\right)}{n^{r-v_1} m^{p-v_2}}$$

$$\left[\left(\frac{\sum_{i_*=h_1}^{v_1} \sum_{j_*=h_2}^{v_2} l_{i_* j_*} \lambda_{1i_*} \lambda_{2j_*}}{h_1! h_2!} \right) \left\| {}^{k_{10}}_{k_{2j}} D^{(0,\alpha_{2j})}_{*(-1)} y^{h_2} \right\|_{\infty,[-1,1]} + \lambda_{2j} \right], \tag{4.53}$$

and

(5) the case of $j = 0$, $i = 1, ..., h_1$ *follows, it holds:*

$$\left\| {}_{k_{20}}^{k_{1i}} D_{*(-1)}^{(\alpha_{1i},0)} f - {}_{k_{20}}^{k_{1i}} D_{*(-1)}^{(\alpha_{1i},0)} Q_{n,m}^* \right\|_{\infty,[-1,1]^2} \leq \frac{\widetilde{C}\omega_{s,q}\left(f^{(r,p)}; \frac{1}{n}, \frac{1}{m}\right)}{n^{r-v_1} m^{p-v_2}}$$

$$\left[\left(\frac{\sum_{i_*=h_1}^{v_1} \sum_{j_*=h_2}^{v_2} l_{i_* j_*} \lambda_{1i_*} \lambda_{2j_*}}{h_1! h_2!} \right) \left\| {}_{k_{20}}^{k_{1i}} D_{*(-1)}^{(\alpha_{1i},0)} x^{h_1} \right\|_{\infty,[-1,1]} + \lambda_{1i} \right]. \tag{4.54}$$

Proof By Corollary 4.3 there exists

$$Q_{n,m} \equiv Q_{n,m}(f) \in (P_n \otimes C([-1,1]) + C([-1,1]) \otimes P_m)$$

such that

$$\left\| f^{(i,j)} - Q_{n,m}^{(i,j)} \right\|_\infty \leq \frac{\widetilde{C}}{n^{r-i} m^{p-j}} \cdot \omega_{s,q}\left(f^{(r,p)}; \frac{1}{n}, \frac{1}{m}\right), \tag{4.55}$$

for all $(0,0) \leq (i,j) \leq (r,p)$, while $Q_{n,m} \in C^{r,p}\left([-1,1]^2\right)$. Here $\widetilde{C}$ depends only on r, p, s, q, where $n \geq \max\{4(r+1), r+s\}$ and $m \geq \max\{4(p+1), p+q\}$, with $r, p \in \mathbb{Z}_+$, $s, q \in \mathbb{N}$, $f \in C^{r,p}\left([-1,1]^2\right)$.

Indeed by [12] we have that $Q_{n,m}^{(r,p)}$ is continuous on $[-1,1]^2$.

We observe the following ($i = 1, ..., r$; $j = 1, ..., p$)

$$\left| {}_{k_{2j}}^{k_{1i}} D_{*(-1)}^{(\alpha_{1i},\alpha_{2j})} f(x_1,x_2) - {}_{k_{2j}}^{k_{1i}} D_{*(-1)}^{(\alpha_{1i},\alpha_{2j})} Q_{n,m}(x_1,x_2) \right| =$$

$$\left| \int_{-1}^{x_1} \int_{-1}^{x_2} k_{1i}(x_1 - t_1) k_{2j}(x_2 - t_2) \frac{\partial^{i+j} f(t_1,t_2)}{\partial t_1^i \partial t_2^j} dt_1 dt_2 - \right.$$

$$\left. \int_{-1}^{x_1} \int_{-1}^{x_2} k_{1i}(x_1 - t_1) k_{2j}(x_2 - t_2) \frac{\partial^{i+j} Q_{n,m}(t_1,t_2)}{\partial t_1^i \partial t_2^j} dt_1 dt_2 \right| = \tag{4.56}$$

$$\left| \int_{-1}^{x_1} \int_{-1}^{x_2} k_{1i}(x_1 - t_1) k_{2j}(x_2 - t_2) \left(\frac{\partial^{i+j} f(t_1,t_2)}{\partial t_1^i \partial t_2^j} - \frac{\partial^{i+j} Q_{n,m}(t_1,t_2)}{\partial t_1^i \partial t_2^j} \right) dt_1 dt_2 \right| \leq$$

$$\int_{-1}^{x_1} \int_{-1}^{x_2} k_{1i}(x_1 - t_1) k_{2j}(x_2 - t_2) \left| \frac{\partial^{i+j} f(t_1,t_2)}{\partial t_1^i \partial t_2^j} - \frac{\partial^{i+j} Q_{n,m}(t_1,t_2)}{\partial t_1^i \partial t_2^j} \right| dt_1 dt_2 \leq$$

$$\left(\int_{-1}^{x_1}\int_{-1}^{x_2} k_{1i}(x_1-t_1)\,k_{2j}(x_2-t_2)\,dt_1dt_2\right)\frac{\widetilde{C}}{n^{r-i}m^{p-j}}\omega_{s,q}\left(f^{(r,p)};\frac{1}{n},\frac{1}{m}\right)=$$

$$\left(\int_{-1}^{x_1} k_{1i}(x_1-t_1)\,dt_1\right)\left(\int_{-1}^{x_2} k_{2j}(x_2-t_2)\,dt_2\right)\frac{\widetilde{C}}{n^{r-i}m^{p-j}}\omega_{s,q}\left(f^{(r,p)};\frac{1}{n},\frac{1}{m}\right)= \tag{4.57}$$

$$\left(\int_{0}^{x_1+1} k_{1i}(z)\,dz\right)\left(\int_{0}^{x_2+1} k_{2j}(z)\,dz\right)\frac{\widetilde{C}}{n^{r-i}m^{p-j}}\omega_{s,q}\left(f^{(r,p)};\frac{1}{n},\frac{1}{m}\right)\leq$$

$$\left(\int_{0}^{2} k_{1i}(z)\,dz\right)\left(\int_{0}^{2} k_{2j}(z)\,dz\right)\frac{\widetilde{C}}{n^{r-i}m^{p-j}}\omega_{s,q}\left(f^{(r,p)};\frac{1}{n},\frac{1}{m}\right).$$

We have proved that

$$\left|{}^{k_{1i}}_{k_{2j}}D_{*(-1)}^{(\alpha_{1i},\alpha_{2j})}f(x_1,x_2)-{}^{k_{1i}}_{k_{2j}}D_{*(-1)}^{(\alpha_{1i},\alpha_{2j})}Q_{n,m}(x_1,x_2)\right|\leq \tag{4.58}$$

$$\left(\int_{0}^{2} k_{1i}(z)\,dz\right)\left(\int_{0}^{2} k_{2j}(z)\,dz\right)\frac{\widetilde{C}}{n^{r-i}m^{p-j}}\omega_{s,q}\left(f^{(r,p)};\frac{1}{n},\frac{1}{m}\right),$$

$\forall\,(x_1,x_2)\in[-1,1]^2;\ i=1,...,r;\ j=1,...,p.$

So we have proved that there exists $Q_{n,m}$ such that

$$\left\|{}^{k_{1i}}_{k_{2j}}D_{*(-1)}^{(\alpha_{1i},\alpha_{2j})}(f)-{}^{k_{1i}}_{k_{2j}}D_{*(-1)}^{(\alpha_{1i},\alpha_{2j})}(Q_{n,m})\right\|_{\infty,[-1,1]^2}\leq \tag{4.59}$$

$$\left(\int_{0}^{2} k_{1i}(z)\,dz\right)\left(\int_{0}^{2} k_{2j}(z)\,dz\right)\frac{\widetilde{C}}{n^{r-i}m^{p-j}}\omega_{s,q}\left(f^{(r,p)};\frac{1}{n},\frac{1}{m}\right),$$

$i=1,...,r;\ j=1,...,p.$

We call

$$\lambda_{1i}:=\int_{0}^{2} k_{1i}(z)\,dz,\quad \lambda_{2j}:=\int_{0}^{2} k_{2j}(z)\,dz, \tag{4.60}$$

for $i=1,...,r;\ j=1,...,p$, as in (4.49).

We also set $\lambda_{10}=\lambda_{20}=1$.

Thus the following inequality is valid in general:

$$\left\| {}_{k_{2j}}^{k_{1i}} D_{*(-1)}^{(\alpha_{1i},\alpha_{2j})} (f) - {}_{k_{2j}}^{k_{1i}} D_{*(-1)}^{(\alpha_{1i},\alpha_{2j})} (Q_{n,m}) \right\|_{\infty,[-1,1]^2} \leq \tag{4.61}$$

$$\lambda_{1i} \lambda_{2j} \frac{\widetilde{C}}{n^{r-i} m^{p-j}} \omega_{s,q} \left(f^{(r,p)}; \frac{1}{n}, \frac{1}{m} \right),$$

for $i = 0, 1, ..., r;\ j = 0, 1, ..., p$.

Define

$$\rho_{n,m} := \widetilde{C} \omega_{s,q} \left(f^{(r,p)}; \frac{1}{n}, \frac{1}{m} \right) \left[\sum_{i=h_1}^{v_1} \sum_{j=h_2}^{v_2} \left(l_{ij} \lambda_{1i} \lambda_{2j} n^{i-r} m^{j-p} \right) \right]. \tag{4.62}$$

(I) Suppose, throughout $[0, 1]^2$, $\alpha_{h_1 h_2}(x, y) \geq \alpha > 0$. Let $Q_{n,m}^*(x, y) \in (P_n \otimes C([-1, 1]) + C([-1, 1]) \otimes P_m)$, $(x, y) \in [-1, 1]^2$, as in (4.61), so that

$$\left\| {}_{k_{2j}}^{k_{1i}} D_{*(-1)}^{(\alpha_{1i},\alpha_{2j})} \left(f(x, y) + \rho_{n,m} \frac{x^{h_1}}{h_1!} \frac{y^{h_2}}{h_2!} \right) - {}_{k_{2j}}^{k_{1i}} D_{*(-1)}^{(\alpha_{1i},\alpha_{2j})} Q_{n,m}^*(x, y) \right\|_{\infty,[-1,1]^2} \leq$$

$$\lambda_{1i} \lambda_{2j} \frac{\widetilde{C}}{n^{r-i} m^{p-j}} \omega_{s,q} \left(f^{(r,p)}; \frac{1}{n}, \frac{1}{m} \right) =: T_{ij}, \tag{4.63}$$

for $i = 0, 1, ..., r;\ j = 0, 1, ..., p$.

If $(h_1 + 1, h_2 + 1) \leq (i, j) \leq (r, p)$, or $0 \leq i \leq h_1$, $h_2 + 1 \leq j \leq p$, or $h_1 + 1 \leq i \leq r, 0 \leq j \leq h_2$ we get from the last

$$\left\| {}_{k_{2j}}^{k_{1i}} D_{*(-1)}^{(\alpha_{1i},\alpha_{2j})} (f) - {}_{k_{2j}}^{k_{1i}} D_{*(-1)}^{(\alpha_{1i},\alpha_{2j})} Q_{n,m}^* \right\|_{\infty,[-1,1]^2} \leq$$

$$\lambda_{1i} \lambda_{2j} \frac{\widetilde{C}}{n^{r-i} m^{p-j}} \omega_{s,q} \left(f^{(r,p)}; \frac{1}{n}, \frac{1}{m} \right), \tag{4.64}$$

proving (4.50).

If $(0, 0) \leq (i, j) \leq (h_1, h_2)$, we get that

$$\left\| {}_{k_{2j}}^{k_{1i}} D_{*(-1)}^{(\alpha_{1i},\alpha_{2j})} f + \frac{\rho_{n,m}}{h_1! h_2!} {}_{k_{2j}}^{k_{1i}} D_{*(-1)}^{(\alpha_{1i},\alpha_{2j})} \left(x^{h_1} y^{h_2} \right) - {}_{k_{2j}}^{k_{1i}} D_{*(-1)}^{(\alpha_{1i},\alpha_{2j})} Q_{n,m}^*(x, y) \right\|_{\infty,[-1,1]^2}$$

$$\leq T_{ij}. \tag{4.65}$$

That is for $(1, 1) \leq (i, j) \leq (h_1, h_2)$, we have

$$\left\| {}_{k_{2j}}^{k_{1i}}D_{*(-1)}^{(\alpha_{1i},\alpha_{2j})} f - {}_{k_{2j}}^{k_{1i}}D_{*(-1)}^{(\alpha_{1i},\alpha_{2j})} Q_{n,m}^{*} \right\|_{\infty,[-1,1]^2} \leq$$

$$\frac{\rho_{n,m}}{h_1!h_2!} \left\| {}_{k_{2j}}^{k_{1i}}D_{*(-1)}^{(\alpha_{1i},\alpha_{2j})} \left(x^{h_1}y^{h_2}\right) \right\|_{\infty,[-1,1]^2} + T_{ij} =$$

$$\frac{\widetilde{C}\omega_{s,q}\left(f^{(r,p)}; \frac{1}{n}, \frac{1}{m}\right)}{h_1!h_2!} \left[\sum_{i_*=h_1}^{v_1} \sum_{j_*=h_2}^{v_2} \left(l_{i_*j_*}\lambda_{1i_*}\lambda_{2j_*}n^{i_*-r}m^{j_*-p}\right) \right]$$

$$\left\| {}_{k_{2j}}^{k_{1i}}D_{*(-1)}^{(\alpha_{1i},\alpha_{2j})} \left(x^{h_1}y^{h_2}\right) \right\|_{\infty,[-1,1]^2} + \lambda_{1i}\lambda_{2j} \frac{\widetilde{C}}{n^{r-i}m^{p-j}} \omega_{s,q}\left(f^{(r,p)}; \frac{1}{n}, \frac{1}{m}\right) =$$

$$\widetilde{C}\omega_{s,q}\left(f^{(r,p)}; \frac{1}{n}, \frac{1}{m}\right) \left\{ \frac{1}{h_1!h_2!} \left[\sum_{i_*=h_1}^{v_1} \sum_{j_*=h_2}^{v_2} l_{i_*j_*}\lambda_{1i_*}\lambda_{2j_*}n^{i_*-r}m^{j_*-p} \right] \right.$$

$$\left. \left\| {}_{k_{2j}}^{k_{1i}}D_{*(-1)}^{(\alpha_{1i},\alpha_{2j})} \left(x^{h_1}y^{h_2}\right) \right\|_{\infty,[-1,1]^2} + \frac{\lambda_{1i}\lambda_{2j}}{n^{r-i}m^{p-j}} \right\} \leq$$

$$\frac{\widetilde{C}\omega_{s,q}\left(f^{(r,p)}; \frac{1}{n}, \frac{1}{m}\right)}{n^{r-v_1}m^{p-v_2}} \left\{ \left(\frac{\sum_{i_*=h_1}^{v_1} \sum_{j_*=h_2}^{v_2} l_{i_*j_*}\lambda_{1i_*}\lambda_{2j_*}}{h_1!h_2!} \right) \right. \tag{4.66}$$

$$\left. \left\| {}_{k_{2j}}^{k_{1i}}D_{*(-1)}^{(\alpha_{1i},\alpha_{2j})} \left(x^{h_1}y^{h_2}\right) \right\|_{\infty,[-1,1]^2} + \lambda_{1i}\lambda_{2j} \right\},$$

proving (4.51).

If $i = j = 0$, from (4.63) we obtain

$$\left\| f + \frac{\rho_{n,m}}{h_1!h_2!} x^{h_1}y^{h_2} - Q_{n,m}^{*} \right\|_{\infty,[-1,1]^2} \leq \frac{\widetilde{C}}{n^r m^p} \omega_{s,q}\left(f^{(r,p)}; \frac{1}{n}, \frac{1}{m}\right), \tag{4.67}$$

and

$$\left\| f - Q_{n,m}^{*} \right\|_{\infty,[-1,1]^2} \leq \frac{\rho_{n,m}}{h_1!h_2!} + \frac{\widetilde{C}}{n^r m^p} \omega_{s,q}\left(f^{(r,p)}; \frac{1}{n}, \frac{1}{m}\right) =$$

$$\frac{\widetilde{C}\omega_{s,q}\left(f^{(r,p)}; \frac{1}{n}, \frac{1}{m}\right)}{h_1!h_2!} \left[\sum_{i_*=h_1}^{v_1} \sum_{j_*=h_2}^{v_2} l_{i_*j_*}\lambda_{1i_*}\lambda_{2j_*}n^{i_*-r}m^{j_*-p} \right]$$

$$+ \frac{\widetilde{C}}{n^r m^p} \omega_{s,q}\left(f^{(r,p)}; \frac{1}{n}, \frac{1}{m}\right) =$$

$$\widetilde{C}\omega_{s,q}\left(f^{(r,p)};\frac{1}{n},\frac{1}{m}\right)\left[\frac{1}{h_1!h_2!}\sum_{i_*=h_1}^{v_1}\sum_{j_*=h_2}^{v_2}l_{i_*j_*}\lambda_{1i_*}\lambda_{2j_*}n^{i_*-r}m^{j_*-p}+\frac{1}{n^r m^p}\right]\leq \tag{4.68}$$

$$\frac{\widetilde{C}\omega_{s,q}\left(f^{(r,p)};\frac{1}{n},\frac{1}{m}\right)}{n^{r-v_1}m^{p-v_2}}\left[\frac{\sum_{i_*=h_1}^{v_1}\sum_{j_*=h_2}^{v_2}l_{i_*j_*}\lambda_{1i_*}\lambda_{2j_*}}{h_1!h_2!}+1\right],$$

proving (4.52).

Next case is for $i=0$, $j=1,\ldots,h_2$, from (4.63) we get

$$\left\|{}_{k_{2j}}^{k_{10}}D_{*(-1)}^{(0,\alpha_{2j})}f+\frac{\rho_{n,m}x^{h_1}}{h_1!h_2!}\,{}_{k_{2j}}^{k_{10}}D_{*(-1)}^{(0,\alpha_{2j})}y^{h_2}-{}_{k_{2j}}^{k_{10}}D_{*(-1)}^{(0,\alpha_{2j})}Q_{n,m}^{*}\right\|_{\infty,[-1,1]^2}\leq \tag{4.69}$$

$$\lambda_{2j}\frac{\widetilde{C}}{n^r m^{p-j}}\omega_{s,q}\left(f^{(r,p)};\frac{1}{n},\frac{1}{m}\right).$$

Therefore we have that

$$\left\|{}_{k_{2j}}^{k_{10}}D_{*(-1)}^{(0,\alpha_{2j})}f-{}_{k_{2j}}^{k_{10}}D_{*(-1)}^{(0,\alpha_{2j})}Q_{n,m}^{*}\right\|_{\infty,[-1,1]^2}\leq$$

$$\frac{\rho_{n,m}}{h_1!h_2!}\left\|{}_{k_{2j}}^{k_{10}}D_{*(-1)}^{(0,\alpha_{2j})}y^{h_2}\right\|_{\infty,[-1,1]^2}+\lambda_{2j}\frac{\widetilde{C}}{n^r m^{p-j}}\omega_{s,q}\left(f^{(r,p)};\frac{1}{n},\frac{1}{m}\right)=$$

$$\frac{\widetilde{C}\omega_{s,q}\left(f^{(r,p)};\frac{1}{n},\frac{1}{m}\right)}{h_1!h_2!}\left[\sum_{i_*=h_1}^{v_1}\sum_{j_*=h_2}^{v_2}\left(l_{i_*j_*}\lambda_{1i_*}\lambda_{2j_*}n^{i_*-r}m^{j_*-p}\right)\right]$$

$$\left\|{}_{k_{2j}}^{k_{10}}D_{*(-1)}^{(0,\alpha_{2j})}y^{h_2}\right\|_{\infty,[-1,1]}+\lambda_{2j}\frac{\widetilde{C}}{n^r m^{p-j}}\omega_{s,q}\left(f^{(r,p)};\frac{1}{n},\frac{1}{m}\right)=$$

$$\widetilde{C}\omega_{s,q}\left(f^{(r,p)};\frac{1}{n},\frac{1}{m}\right)\left[\frac{\left[\sum_{i_*=h_1}^{v_1}\sum_{j_*=h_2}^{v_2}\left(l_{i_*j_*}\lambda_{1i_*}\lambda_{2j_*}n^{i_*-r}m^{j_*-p}\right)\right]}{h_1!h_2!}\right.$$

$$\left.\left\|{}_{k_{2j}}^{k_{10}}D_{*(-1)}^{(0,\alpha_{2j})}y^{h_2}\right\|_{\infty,[-1,1]}+\frac{\lambda_{2j}}{n^r m^{p-j}}\right]\leq \tag{4.70}$$

$$\frac{\widetilde{C}\omega_{s,q}\left(f^{(r,p)};\frac{1}{n},\frac{1}{m}\right)}{n^{r-v_1}m^{p-v_2}}\left[\frac{\left(\sum_{i_*=h_1}^{v_1}\sum_{j_*=h_2}^{v_2}l_{i_*j_*}\lambda_{1i_*}\lambda_{2j_*}\right)}{h_1!h_2!}\right.$$

$$\left.\left\|{}_{k_{2j}}^{k_{10}}D_{*(-1)}^{(0,\alpha_{2j})}y^{h_2}\right\|_{\infty,[-1,1]}+\lambda_{2j}\right],$$

proving (4.53).

The case of $j = 0$, $i = 1, ..., h_1$, is met similarly as in (4.70). Namely we get

$$\left\| {}_{k_{20}}^{k_{1i}} D_{*(-1)}^{(\alpha_{1i},0)} f - {}_{k_{20}}^{k_{1i}} D_{*(-1)}^{(\alpha_{1i},0)} Q_{n,m}^{*} \right\|_{\infty,[-1,1]^2} \leq$$

$$\frac{\widetilde{C}\omega_{s,q}\left(f^{(r,p)}; \frac{1}{n}, \frac{1}{m}\right)}{n^{r-v_1} m^{p-v_2}} \left[\left(\frac{\sum_{i_*=h_1}^{v_1} \sum_{j_*=h_2}^{v_2} l_{i_* j_*} \lambda_{1i_*} \lambda_{2j_*}}{h_1! h_2!} \right) \right. \tag{4.71}$$

$$\left. \left\| {}_{k_{20}}^{k_{1i}} D_{*(-1)}^{(\alpha_{1i},0)} x^{h_1} \right\|_{\infty,[-1,1]} + \lambda_{1i} \right],$$

proving (4.54).

So if $(x, y) \in [0, 1]^2$, we can write

$$\alpha_{h_1 h_2}^{-1}(x, y) L^*\left(Q_{n,m}^*(x, y)\right) = \alpha_{h_1 h_2}^{-1}(x, y) L^*(f(x, y)) +$$

$$\frac{\rho_{n,m}}{h_1! h_2!} {}_{k_{2h_2}}^{k_{1h_1}} D_{*(-1)}^{(\alpha_{1h_1},\alpha_{2h_2})}\left(x^{h_1} y^{h_2}\right) + \sum_{i=h_1}^{v_1} \sum_{j=h_2}^{v_2} \alpha_{h_1 h_2}^{-1}(x, y) \alpha_{ij}(x, y) \cdot$$

$$\left[{}_{k_{2j}}^{k_{1i}} D_{*(-1)}^{(\alpha_{1i},\alpha_{2j})} Q_{n,m}^*(x, y) - {}_{k_{2j}}^{k_{1i}} D_{*(-1)}^{(\alpha_{1i},\alpha_{2j})} f(x, y) - \frac{\rho_{n,m}}{h_1! h_2!} {}_{k_{2j}}^{k_{1i}} D_{*(-1)}^{(\alpha_{1i},\alpha_{2j})}\left(x^{h_1} y^{h_2}\right) \right] \tag{4.72}$$

(by $L^* f \geq 0$)

$$\overset{(4.63)}{\geq} \frac{\rho_{n,m}}{h_1! h_2!} {}_{k_{2h_2}}^{k_{1h_1}} D_{*(-1)}^{(\alpha_{1h_1},\alpha_{2h_2})}\left(x^{h_1} y^{h_2}\right) -$$

$$\left(\sum_{i=h_1}^{v_1} \sum_{j=h_2}^{v_2} l_{ij} \lambda_{1i} \lambda_{2j} n^{i-r} m^{j-p} \right) \widetilde{C}\omega_{s,q}\left(f^{(r,p)}; \frac{1}{n}, \frac{1}{m}\right) =$$

$$\rho_{n,m} \left(\frac{{}_{k_{2h_2}}^{k_{1h_1}} D_{*(-1)}^{(\alpha_{1h_1},\alpha_{2h_2})}\left(x^{h_1} y^{h_2}\right)}{h_1! h_2!} - 1 \right) = \tag{4.73}$$

$$\rho_{n,m} \left[\frac{\left({}_{k_{20}}^{k_{1h_1}} D_{*(-1)}^{(\alpha_{1h_1},0)} x^{h_1} \right) \left({}_{k_{2h_2}}^{k_{10}} D_{*(-1)}^{(0,\alpha_{2h_2})} y^{h_2} \right)}{h_1! h_2!} - 1 \right] =: \varphi.$$

If $h_1 = h_2 = 0$, then $\alpha_{1h_1} = \alpha_{2h_2} = 0$, and $\varphi = 0$.

If $h_1 = 0$, and $h_2 \neq 0$, then

$$\varphi = \rho_{n,m} \left[\frac{{}_{k_{2h_2}}^{k_{10}} D_{*(-1)}^{(0,\alpha_{2h_2})} y^{h_2}}{h_2!} - 1 \right] =$$

$$\rho_{n,m} \left[\int_{-1}^{y} k_{2h_2} (y-t)\, dt - 1 \right] = \rho_{n,m} \left(\int_{0}^{y+1} k_{2h_2} (z)\, dz - 1 \right) \geq \tag{4.74}$$

$$\rho_{n,m} \left(\int_{0}^{1} k_{2h_2} (z)\, dz - 1 \right) \geq 0,$$

by the assumption (4.40).

Similarly, we treat the case $h_1 \neq 0, h_2 = 0$.

When $h_1, h_2 \neq 0$, then we have

$$\varphi = \rho_{n,m} \left[\left(\int_{-1}^{x} k_{1h_1} (x-t)\, dt \right) \left(\int_{-1}^{y} k_{2h_2} (y-t)\, dt \right) - 1 \right] = \tag{4.75}$$

$$\rho_{n,m} \left[\left(\int_{0}^{x+1} k_{1h_1} (z)\, dz \right) \left(\int_{0}^{y+1} k_{2h_2} (z)\, dz \right) - 1 \right] \geq$$

$$\rho_{n,m} \left[\left(\int_{0}^{1} k_{1h_1} (z)\, dz \right) \left(\int_{0}^{1} k_{2h_2} (z)\, dz \right) - 1 \right] \geq 0,$$

by the assumption (4.40).

So in all four cases we get that

$$L^* \left(Q_{n,m}^* (x,y) \right) \geq 0, \ \forall \ (x,y) \in [0,1]^2 . \tag{4.76}$$

(II) Suppose, throughout $[0,1]^2$, $\alpha_{h_1 h_2} (x,y) \leq \beta < 0$. Let $Q_{n,m}^* (x,y) \in (P_n \otimes C([-1,1]) + C([-1,1]) \otimes P_m)$, $(x,y) \in [-1,1]^2$, as in (4.61), so that

$$\left\| {}_{k_{2j}}^{k_{1i}} D_{*(-1)}^{(\alpha_{1i},\alpha_{2j})} \left(f(x,y) - \rho_{n,m} \frac{x^{h_1}}{h_1!} \frac{y^{h_2}}{h_2!} \right) - {}_{k_{2j}}^{k_{1i}} D_{*(-1)}^{(\alpha_{1i},\alpha_{2j})} Q_{n,m}^{**} (x,y) \right\|_{\infty,[-1,1]^2} \leq$$

$$\lambda_{1i} \lambda_{2j} \frac{\widetilde{C}}{n^{r-i} m^{p-j}} \omega_{s,q} \left(f^{(r,p)}; \frac{1}{n}, \frac{1}{m} \right), \tag{4.77}$$

for $i = 0, 1, \ldots, r$, $j = 0, 1, \ldots, p$.

Similarly, we get as in the first case $\geq \alpha > 0$, the inequalities of simultaneous fractional convergence, see {(4.51), (4.66)}, {(4.52), (4.68)}, {(4.53),(4.70)}, {(4.54), (4.71)}.

So if $(x, y) \in [0, 1]^2$ we can write

$$\alpha_{h_1 h_2}^{-1}(x, y) L^* \left(Q_{n,m}^*(x, y)\right) = \alpha_{h_1 h_2}^{-1}(x, y) L^*(f(x, y)) -$$

$$\frac{\rho_{n,m}}{h_1! h_2!} {}_{k_{2h_2}}^{k_{1h_1}} D_{*(-1)}^{(\alpha_{1h_1}, \alpha_{2h_2})} \left(x^{h_1} y^{h_2}\right) + \sum_{i=h_1}^{v_1} \sum_{j=h_2}^{v_2} \alpha_{h_1 h_2}^{-1}(x, y) \alpha_{ij}(x, y) \cdot \tag{4.78}$$

$$\left[{}_{k_{2j}}^{k_{1i}} D_{*(-1)}^{(\alpha_{1i}, \alpha_{2j})} Q_{n,m}^*(x, y) - {}_{k_{2j}}^{k_{1i}} D_{*(-1)}^{(\alpha_{1i}, \alpha_{2j})} f(x, y) + \frac{\rho_{n,m}}{h_1! h_2!} {}_{k_{2j}}^{k_{1i}} D_{*(-1)}^{(\alpha_{1i}, \alpha_{2j})} \left(x^{h_1} y^{h_2}\right) \right]$$

(by $L^* f \geq 0$)

$$\overset{(4.78)}{\leq} -\frac{\rho_{n,m}}{h_1! h_2!} {}_{k_{2h_2}}^{k_{1h_1}} D_{*(-1)}^{(\alpha_{1h_1}, \alpha_{2h_2})} \left(x^{h_1} y^{h_2}\right) +$$

$$\left(\sum_{i=h_1}^{v_1} \sum_{j=h_2}^{v_2} l_{ij} \lambda_{1i} \lambda_{2j} n^{r-i} m^{p-j} \right) \widetilde{C} \omega_{s,q} \left(f^{(r,p)}; \frac{1}{n}, \frac{1}{m} \right) = \tag{4.79}$$

$$\rho_{n,m} \left[1 - \frac{{}_{k_{2h_2}}^{k_{1h_1}} D_{*(-1)}^{(\alpha_{1h_1}, \alpha_{2h_2})} \left(x^{h_1} y^{h_2}\right)}{h_1! h_2!} \right] =$$

$$\rho_{n,m} \left[1 - \frac{\left({}_{k_{20}}^{k_{1h_1}} D_{*(-1)}^{(\alpha_{1h_1}, 0)} x^{h_1} \right) \left({}_{k_{2h_2}}^{k_{10}} D_{*(-1)}^{(0, \alpha_{2h_2})} y^{h_2} \right)}{h_1! h_2!} \right] =: \psi.$$

If $h_1 = h_2 = 0$, then $\alpha_{1h_1} = \alpha_{2h_2} = 0$, and $\psi = 0$.

If $h_1 = 0$, and $h_2 \neq 0$, then

$$\psi = \rho_{n,m} \left[1 - \frac{{}_{k_{2h_2}}^{k_{10}} D_{*(-1)}^{(0, \alpha_{2h_2})} y^{h_2}}{h_2!} \right] = \rho_{n,m} \left[1 - \int_{-1}^{y} k_{2h_2}(y - t) \, dt \right] = \tag{4.80}$$

$$\rho_{n,m} \left[1 - \int_{0}^{y+1} k_{2h_2}(z) \, dz \right] \leq \rho_{n,m} \left[1 - \int_{0}^{1} k_{2h_2}(z) \, dz \right] \leq 0,$$

see (4.40).

Similarly, we treat the case $h_1 \neq 0$, $h_2 = 0$.

When $h_1, h_2 \neq 0$, then we have

$$\psi = \rho_{n,m}\left[1-\left(\int_{-1}^{x} k_{1h_1}(x-t)\,dt\right)\left(\int_{-1}^{y} k_{2h_2}(y-t)\,dt\right)\right] =$$

$$\rho_{n,m}\left[1-\left(\int_{0}^{x+1} k_{1h_1}(z)\,dz\right)\left(\int_{0}^{y+1} k_{2h_2}(z)\,dz\right)\right] \leq \tag{4.81}$$

$$\rho_{n,m}\left[1-\left(\int_{0}^{1} k_{1h_1}(z)\,dz\right)\left(\int_{0}^{1} k_{2h_2}(z)\,dz\right)\right] \leq 0,$$

by the assumption (4.40).

So in all four cases we proved again that

$$L^*\left(Q^*_{n,m}(x,y)\right) \geq 0, \ \forall\ (x,y) \in [0,1]^2. \tag{4.82}$$

The proof is complete.

Conclusion 4.11 Clearly Theorem 4.10 generalizes Theorem 4.8 to many fractional calculi, opening new avenues of fractional research activity. The approximating pseudo-polynomial $Q^*_{n,m}$ depends on f, $\rho_{n,m}$, h_1, h_2; which $\rho_{n,m}$ depends on $\widetilde{C}$ (which depends on r, p, s, q), f, n, m, l_{ij}, λ_{1i}, λ_{2j}; and which: λ_{1i} depends on k_{1i} and λ_{2j} depends on k_{2j}. That is $Q^*_{n,m}$ depends on the type of bivariate fractional calculus we use.

Consequently, Theorem 4.10 is valid at least for the following important bivariate left fractional linear differential operators:

(1)

$$L_1^* := \sum_{i=h_1}^{v_1}\sum_{j=h_2}^{v_2} \alpha_{ij}(x,y)\left[{}^{C}D^{(\gamma_1,\gamma_2)}_{(\rho_1,\rho_2),(\alpha_{1i},\alpha_{2j}),(\omega_1,\omega_2),(-1+,-1+)}\right], \tag{4.83}$$

where $\rho_1, \rho_2 > 0, \gamma_1, \gamma_2 < 0$, and $\omega_1, \omega_2 > 0$ large enough (from bivariate Prabhakar fractional calculus, see (4.42));

(2)

$$L_2^* := \sum_{i=h_1}^{v_1}\sum_{j=h_2}^{v_2} \alpha_{ij}(x,y)\left[D^{(\alpha_{1i},\alpha_{2j})}_{-1*}\right], \tag{4.84}$$

(see (4.45)) where $\theta = 1, 2$; $\gamma_{\theta j} > 0$, $j = 1, ..., m$; $\lambda_\theta = 1$, $0 < \rho_\theta < 1$; and small enough $\omega_{\theta j} < 0$, $j = 1, ..., m$ (from bivariate generalized non-singular fractional calculus);

and
(3)

$$L_3^* := \sum_{i=h_1}^{v_1} \sum_{j=h_2}^{v_2} \alpha_{ij}(x, y) \left[{}_{(\omega_1,\omega_2)}^{CF} D_{-1+}^{(\alpha_{1i},\alpha_{2j})} \right], \tag{4.85}$$

(see (4.46)) for small enough $\omega_1, \omega_2 < 0$ (from bivariate parametrized Caputo-Fabrizio non-singular kernel fractional calculus).

Our developed bivariate abstract fractional monotone approximation theory by pseudopolynomials with its applications, involves weaker conditions than the one with ordinary partial derivatives, see Theorem 4.4, and can manage many diverse general cases in a multitude of complex settings and environments.

References

1. Anastassiou, G.A.: Monotone Approximation by Pseudopolynomials, Approximation Theory, pp. 5–11. Academic Press, New York (1991)
2. Anastassiou, G.A.: Bivariate Monotone Approximation. Proc. Amer. Math. Soc. **112**(4), 959–964 (1991)
3. Anastassiou, G.A.: Bivariate Left Fractional Pseudo-polynomial Monotone Approximation, Computational Analysis, pp. 1–17. AMAT, Ankara, May 2015, Springer, New York (2016)
4. Anastassiou, G.A.: Multiparameter Fractional Differentiation with Non Singular Kernel. Issues of Analysis, accepted (2021)
5. Anastassiou, G.A.: Foundations of Generalized Prabhakar-Hilfer Fractional Calculus with Applications. Cubo, accepted (2021)
6. Anastassiou, G.A.: Univariate Simultaneous High Order Abstract Fractional Monotone Approximation with Applications. Submitted (2021)
7. Anastassiou, G.A.: Abstract Bivariate Left Fractional Pseudo-Polynomial Monotone Constrained Approximation with Applications. Submitted (2021)
8. Anastassiou, G.A., Shisha, O.: Monotone approximation with linear differential operators. J. Approx. Theory **44**, 391–393 (1985)
9. Atangana, A., Baleanu, D.: New fractional derivatives with nonlocal and non-singular kernel: theory and application to heat transfer model. Therm. Sci. **20**(2), 763–769 (2016)
10. Giusti, A., et al.: A practical Guide to Prabhakar Fractional Calculus. Fractional Calculus Appl. Anal. **23**(1), 9–54 (2020)
11. Gonska, H.H., Simultaneously approximation by algebraic blending functions, Alfred Haar Memorial Conference, Budapest,: Coloquia Mathematica Soc. Janos Bolyai, 49. North-Holand, Amsterdam, 363–382 (1985)
12. Gonska, H.H.: Personal communication with author, 2-24-2014
13. Gorenflo, R., Kilbas, A., Mainardi, F., Rogosin, S.: Mittag-Leffler functions. Related Topics and Applications, Springer, Heidelberg, New York (2014)
14. Losada, J., Nieto, J.J.: Properties of a new fractional derivative without singular Kernel. Progr. Fract. Differ. Appl. **1**(2), 87–92 (2015)
15. Mamatov, T., Samko, S.: Mixed fractional integration operators in mixed weighted Hölder spaces. Fractional Calculus and Appl. Anal. **13**(3), 245–259 (2010)
16. Marchaud, A.: Differences et deerivees d'une fonction de deux variables. C.R. Acad. Sci. **178**, 1467–1470 (1924)
17. Marchaud, A.: Sur les derivees et sur les differences des fonctions de variables reelles. J. Math. Pures Appl. **6**, 337–425 (1927)

18. Polito, F., Tomovski, Z.: Some properties of Prabhakar-type fractional calculus operators. Fractional Differ. Calculus **6**(1), 73–94 (2016)
19. Saxena, R.K., Kalla, S.L.: Ravi Saxena. Multivariate analogue of generalized Mittag-Leffler function. Integr. Transforms Special Funct. **22**(7), 533–548 (2011)
20. Schumaker, L.L.: Spline Functions: Basic Theory. Wiley, New York (1981)
21. Shisha, O.: Monotone approximation. Pac. J. Math. **15**, 667–671 (1965)
22. Srivastava, H.M., Daoust, M.C.: A note on the convergence of Kompe' de Feriet's double hypergeometric series. Math. Nachr. **53**, 151–159 (1972)

Chapter 5
Right Side Abstract Bivariate Monotone Constrained Approximation by Pseudo-polynomials

Here we extend our earlier right side bivariate high order simultaneous fractional monotone constrained approximation theory by pseudo-polynomials to right side abstract bivariate high order simultaneous fractional monotone constrained approximation by pseudo-polynomials, with applications to right side bivariate Prabhakar fractional calculus and non-singular kernel fractional calculi.

So we deal with the following general two-dimensional problem: Let f be a two variable continuously differentiable real valued function of a given order, let L^* be a linear right side abstract fractional mixed partial differential operator and suppose that $L^*(f) \geq 0$ on the critical region of $[-1, 0]^2$. Then for specific and sufficiently large $n, m \in \mathbb{N}$, we can find a sequence of pseudo-polynomials $Q^*_{n,m}$ in two variables with the property $L^*\left(Q^*_{n,m}\right) \geq 0$ on this critical region of $[-1, 0]^2$ such that f is approximated with rates fractionally and simultaneously by $Q^*_{n,m}$ in the uniform norm on the whole domain of f. This constrained approximation is given via inequalities involving the mixed modulus of smoothness $\omega_{s,q}$, $s, q \in \mathbb{N}$, of highest order integer partial derivative of f. It follows [7].

5.1 Introduction

The topic of monotone approximation theory started in [21] and it has become a major trend of approximation theory. A typical problem in this subject is: given a positive integer k, approximate a given function whose kth derivative is ≥ 0 by polynomials having this property.

In [8] the authors replaced the kth derivative with a linear differential operator of order k. We mention this motivating result.

G. A. Anastassiou, *Abstract Fractional Monotone Approximation, Theory and Applications*, Studies in Systems, Decision and Control 411,
https://doi.org/10.1007/978-3-030-95943-2_5

Theorem 5.1 *Let h, k, p be integers, $0 \leq h \leq k \leq p$ and let f be a real function, $f^{(p)}$ continuous in $[-1, 1]$ with modulus of continuity $\omega_1 \left(f^{(p)}, x\right)$ there. Let $a_j(x)$, $j = h, h+1, ..., k$ be real functions, defined and bounded on $[-1, 1]$ and assume $a_h(x)$ is either $\geq$ some number $\alpha > 0$ or $\leq$ some number $\beta < 0$ throughout $[-1, 1]$. Consider the operator*

$$L = \sum_{j=h}^{k} a_j(x) \left[\frac{d^j}{dx^j}\right] \tag{5.1}$$

and suppose, throughout $[-1, 1]$,

$$L(f) \geq 0. \tag{5.2}$$

Then, for every integer $n \geq 1$, there is a real polynomial $Q_n(x)$ of degree $\leq n$ such that

$$L(Q_n) \geq 0 \text{ throughout } [-1, 1] \tag{5.3}$$

and

$$\max_{-1 \leq x \leq 1} |f(x) - Q_n(x)| \leq C n^{k-p} \omega_1 \left(f^{(p)}, \frac{1}{n}\right), \tag{5.4}$$

where C is independent of n or f.

Next let $n, m \in \mathbb{Z}_+$, P_θ denote the space of algebraic polynomials of degree $\leq \theta$. Consider the tensor product spaces $P_n \otimes C([-1, 1])$, $C([-1, 1]) \otimes P_m$ and their sum $P_n \otimes C([-1, 1]) + C([-1, 1]) \otimes P_m$, that is

$$P_n \otimes C([-1, 1]) + C([-1, 1]) \otimes P_m =$$

$$\left\{\sum_{i=0}^{n} x^i A_i(y) + \sum_{j=0}^{m} B_j(x) y^j;\ A_i, B_j \in C([-1, 1]),\ x, y \in [-1, 1]\right\}. \tag{5.5}$$

This is the space of pseudo-polynomials of degree $\leq (n, m)$, first introduced by A. Marchaud in 1924-1927 (see [16, 17]). Here $f^{(k,l)}$ denotes $\frac{\partial^{k+l} f}{\partial x^k \partial y^l}$, the (k, l)-partial derivative of f.

Here we consider the space $C^{r,p}\left([-1, 1]^2\right) = \{f : [-1, 1]^2 \to \mathbb{R}; f^{(k,l)}$ is continuous for $0 \leq k \leq r, 0 \leq l \leq p\}$. Let $f \in C\left([-1, 1]^2\right)$; for $\delta_1, \delta_2 \geq 0$, define the mixed modulus of smoothness of order (s, q), $s, q \in \mathbb{N}$ (see [20, pp. 516–517]) by

$$\omega_{s,q}(f; \delta_1, \delta_2) \equiv \sup\left\{\left|{}_x\Delta_{h_1}^s \circ_y \Delta_{h_2}^q f(x, y)\right| : (x, y),\right. \tag{5.6}$$

$$\left.(x + sh_1, y + qh_2) \in [-1, 1]^2, |h_i| \leq \delta_i, i = 1, 2\right\}.$$

Here

$$ {}_x\Delta_{h_1}^s \circ_y \Delta_{h_2}^q f(x,y) \equiv \sum_{\sigma=0}^{s}\sum_{\mu=0}^{q}(-1)^{s+q-\sigma-\mu}\binom{s}{\sigma}\binom{q}{\mu} f(x+\sigma h_1, y+\mu h_2) \tag{5.7} $$

is a mixed difference of order (s, q).

We mention

Theorem 5.2 (Gonska [11]). *Let $r, p \in \mathbb{Z}_+$, $s, q \in \mathbb{N}$, and $f \in C^{r,p}\left([-1,1]^2\right)$. Let $n, m \in \mathbb{N}$ with $n \geq \max\{4(r+1), r+s\}$ and $m \geq \max\{4(p+1), p+q\}$. Then there exists a linear operator $Q_{n,m}$ from $C^{r,p}\left([-1,1]^2\right)$ into the space of pseudopolynomials $(P_n \otimes C([-1,1]) + C([-1,1]) \otimes P_m)$ such that*

$$ \left|\left(f - Q_{n,m}(f)\right)^{(k,l)}(x,y)\right| \leq \tag{5.8} $$

$$ M_{r,s} \cdot M_{p,q}\left(\Delta_n(x)\right)^{r-k} \cdot \left(\Delta_m(y)\right)^{p-l} \cdot \omega_{s,q}\left(f^{(r,p)}; \Delta_n(x), \Delta_m(y)\right), $$

for all $(0,0) \leq (k,l) \leq (r,p)$, $x, y \in [-1,1]$, where

$$ \Delta_\theta(z) = \frac{\sqrt{1-z^2}}{\theta} + \frac{1}{\theta^2}, \ \theta = n, m; \ z = x, y \in [-1,1]. \tag{5.9} $$

The constants $M_{r,s}$, $M_{p,q}$, are independent of f, (x,y) and (n,m); they depend only on (r,s), (p,q), respectively.

See also [12], saying that $Q_{n,m}^{(r,p)}(f)$ is continuous on $[-1,1]^2$.

We need the following result which is an easy consequence of the last theorem (see [20, p. 517]).

Corollary 5.3 ([1]) *Let $r, p \in \mathbb{Z}_+$, $s, q \in \mathbb{N}$, and $f \in C^{r,p}\left([-1,1]^2\right)$. Let $n, m \in \mathbb{N}$ with $n \geq \max\{4(r+1), r+s\}$ and $m \geq \max\{4(p+1), p+q\}$. Then there exists a pseudopolynomial $Q_{n,m} \equiv Q_{n,m}(f) \in (P_n \otimes C([-1,1]) + C([-1,1]) \otimes P_m)$ such that*

$$ \left\|f^{(k,l)} - Q_{n,m}^{(k,l)}\right\|_\infty \leq \frac{\widetilde{C}}{n^{r-k}m^{p-l}} \cdot \omega_{s,q}\left(f^{(r,p)}; \frac{1}{n}, \frac{1}{m}\right), \tag{5.10} $$

for all $(0,0) \leq (k,l) \leq (r,p)$. Here the constant $\widetilde{C}$ depends only on r, p, s, q.

Corollary 5.3 was used in the proof of the motivational result that follows.

Theorem 5.4 ([1]) *Let h_1, h_2, v_1, v_2, r, p be integers, $0 \leq h_1 \leq v_1 \leq r$, $0 \leq h_2 \leq v_2 \leq p$ and let $f \in C^{r,p}\left([-1,1]^2\right)$, with $f^{(r,p)}$ having a mixed modulus of smoothness $\omega_{s,q}\left(f^{(r,p)}; x, y\right)$ there, $s, q \in \mathbb{N}$. Let $\alpha_{i,j}(x,y)$, $i = h_1, h_1+1, ..., v_1$; $j = h_2, h_2+1, ..., v_2$ be real-valued functions, defined and bounded in $[-1,1]^2$ and suppose $\alpha_{h_1h_2}$ is either $\geq \alpha > 0$ or $\leq \beta < 0$ throughout $[-1,1]^2$. Take the operator*

$$L = \sum_{i=h_1}^{v_1} \sum_{j=h_2}^{v_2} \alpha_{ij}(x, y) \frac{\partial^{i+j}}{\partial x^i \partial y^j} \tag{5.11}$$

and assume, throughout $[-1, 1]^2$ *that*

$$L(f) \geq 0. \tag{5.12}$$

Then for any integers n, m *with* $n \geq \max\{4(r+1), r+s\}$, $m \geq \max\{4(p+1), p+q\}$, *there exists a pseudopolynomial*

$$Q_{n,m} \in (P_n \otimes C([-1, 1]) + C([-1, 1]) \otimes P_m)$$

such that $L(Q_{m,n}) \geq 0$ *throughout* $[-1, 1]^2$ *and*

$$\left\| f^{(k,l)} - Q_{n,m}^{(k,l)} \right\|_\infty \leq \frac{C}{n^{r-v_1} m^{p-v_2}} \cdot \omega_{s,q}\left(f^{(r,p)}; \frac{1}{n}, \frac{1}{m}\right), \tag{5.13}$$

for all $(0, 0) \leq (k, l) \leq (h_1, h_2)$. *Moreover we get*

$$\left\| f^{(k,l)} - Q_{n,m}^{(k,l)} \right\|_\infty \leq \frac{C}{n^{r-k} m^{p-l}} \cdot \omega_{s,q}\left(f^{(r,p)}; \frac{1}{n}, \frac{1}{m}\right), \tag{5.14}$$

for all $(h_1+1, h_2+1) \leq (k, l) \leq (r, p)$. *Also (5.14) is valid whenever* $0 \leq k \leq h_1$, $h_2+1 \leq l \leq p$ *or* $h_1+1 \leq k \leq r$, $0 \leq l \leq h_2$. *Here* C *is a constant independent of* f *and* n, m. *It depends only on* r, p, s, q, L.

We are also motivated by [2].

We mention

Definition 5.5 (*see* [14]) Let $[-1, 1]^2$; $\alpha_1, \alpha_2 > 0$; $\alpha = (\alpha_1, \alpha_2)$, $f \in C([-1, 1]^2)$, $x = (x_1, x_2)$, $t = (t_1, t_2) \in [-1, 1]^2$. We define the right mixed Riemann-Liouville fractional two dimensional integral of order α

$$\left(I_{1-}^{\alpha} f\right)(x) := \frac{1}{\Gamma(\alpha_1)\Gamma(\alpha_2)} \int_{x_1}^{1} \int_{x_2}^{1} (t_1 - x_1)^{\alpha_1 - 1} (t_2 - x_2)^{\alpha_2 - 1} f(t_1, t_2)\, dt_1 dt_2, \tag{5.15}$$

with $x_1, x_2 < 1$.

Notice here that $I_{1-}^{\alpha}(|f|) < \infty$.

Definition 5.6 ([3, pp. 15–31]) Let $\alpha_1, \alpha_2 > 0$ with $\lceil \alpha_1 \rceil = m_1$, $\lceil \alpha_2 \rceil = m_2$, ($\lceil \cdot \rceil$ ceiling of the number). Let here $f \in C^{m_1, m_2}([-1, 1]^2)$. We consider the right Caputo type fractional partial derivative:

$$D_{1-}^{(\alpha_1, \alpha_2)} f(x) := \frac{(-1)^{m_1 + m_2}}{\Gamma(m_1 - \alpha_1)\Gamma(m_2 - \alpha_2)} \cdot$$

$$\int_{x_1}^{1}\int_{x_2}^{1}(t_1-x_1)^{m_1-\alpha_1-1}(t_2-x_2)^{m_2-\alpha_2-1}\frac{\partial^{m_1+m_2}f(t_1,t_2)}{\partial t_1^{m_1}\partial t_2^{m_2}}dt_1dt_2, \tag{5.16}$$

$\forall x=(x_1,x_2)\in[-1,1]^2$, where Γ is the gamma function

$$\Gamma(\nu)=\int_0^{\infty}e^{-t}t^{\nu-1}dt,\quad \nu>0. \tag{5.17}$$

We set

$$D_{1-}^{(0,0)}f(x):=f(x),\quad \forall x\in[-1,1]^2; \tag{5.18}$$

$$D_{1-}^{(m_1,m_2)}f(x):=(-1)^{m_1+m_2}\frac{\partial^{m_1+m_2}f(x)}{\partial x_1^{m_1}\partial x_2^{m_2}},\quad \forall x\in[-1,1]^2. \tag{5.19}$$

Definition 5.7 ([3], pp. 15-31) We also set

$$D_{1-}^{(0,\alpha_2)}f(x):=\frac{(-1)^{m_2}}{\Gamma(m_2-\alpha_2)}\int_{x_2}^{1}(t_2-x_2)^{m_2-\alpha_2-1}\frac{\partial^{m_2}f(x_1,t_2)}{\partial t_2^{m_2}}dt_2, \tag{5.20}$$

$$D_{1-}^{(\alpha_1,0)}f(x):=\frac{(-1)^{m_1}}{\Gamma(m_1-\alpha_1)}\int_{x_1}^{1}(t_1-x_1)^{m_1-\alpha_1-1}\frac{\partial^{m_1}f(t_1,x_2)}{\partial t_1^{m_1}}dt_1, \tag{5.21}$$

and

$$D_{1-}^{(m_1,\alpha_2)}f(x):=\frac{(-1)^{m_2}}{\Gamma(m_2-\alpha_2)}\int_{x_2}^{1}(t_2-x_2)^{m_2-\alpha_2-1}\frac{\partial^{m_1+m_2}f(x_1,t_2)}{\partial x_1^{m_1}\partial t_2^{m_2}}dt_2, \tag{5.22}$$

$$D_{1-}^{(\alpha_1,m_2)}f(x):=\frac{(-1)^{m_1}}{\Gamma(m_1-\alpha_1)}\int_{x_1}^{1}(t_1-x_1)^{m_1-\alpha_1-1}\frac{\partial^{m_1+m_2}f(t_1,x_2)}{\partial t_1^{m_1}\partial x_2^{m_2}}dt_1. \tag{5.23}$$

In [3, pp. 15–31], we extended Theorem 5.4 to the fractional level. Indeed there L is replaced by $\overline{L}$, a linear right Caputo fractional mixed partial differential operator. Now the monotonicity property holds true only on the critical square of $[-1,0]^2$. Simultaneously fractional convergence remains true on all of $[-1,1]^2$.

So we have proved there

Theorem 5.8 *Let h_1, h_2, v_1, v_2, r, p be integers, $0\le h_1\le v_1\le r$, $0\le h_2\le v_2\le p$ and let $f\in C^{r,p}\left([-1,1]^2\right)$, with $f^{(r,p)}$ having a mixed modulus of smoothness*

$\omega_{s,q}\left(f^{(r,p)};x,y\right)$ *there,* $s,q\in\mathbb{N}$. *Let* $\alpha_{ij}(x,y)$, $i=h_1,h_1+1,...,v_1$; $j=h_2,h_2+1,...,v_2$ *be real valued functions, defined and bounded in* $[-1,1]^2$ *and suppose* $\alpha_{h_1h_2}$ *is either* $\geq\alpha>0$ *or* $\leq\beta<0$ *throughout* $[-1,0]^2$. *Assume that* $h_1+h_2=2\gamma$, $\gamma\in\mathbb{Z}_+$. *Here* $n,m\in\mathbb{N}:n\geq\max\{4(r+1),r+s\}$, $m\geq\max\{4(p+1),p+q\}$. *Set*

$$l_{ij}:=\sup_{(x,y)\in[-1,1]^2}\left|\alpha_{h_1h_2}^{-1}(x,y)\,\alpha_{ij}(x,y)\right|<\infty, \tag{5.24}$$

for all $h_1\leq i\leq v_1$, $h_2\leq j\leq v_2$. *Let* $\alpha_{1i},\alpha_{2j}\geq0$ *with* $\lceil\alpha_{1i}\rceil=i$, $\lceil\alpha_{2j}\rceil=j$, $i=0,1,...,r$; $j=0,1,...,p$, *(*$\lceil\cdot\rceil$ *ceiling of the number),* $\alpha_{10}=0$, $\alpha_{20}=0$.

Consider the right fractional bivariate differential operator

$$\overline{L}:=\sum_{i=h_1}^{v_1}\sum_{j=h_2}^{v_2}\alpha_{ij}(x,y)\,D_{1-}^{(\alpha_{1i},\alpha_{2j})}. \tag{5.25}$$

Assume $\overline{L}f(x,y)\geq0$, *on* $[-1,0]^2$.

Then there exists

$$Q_{n,m}^*\equiv Q_{n,m}^*(f)\in(P_n\otimes C([-1,1])+C([-1,1])\otimes P_m)$$

such that $\overline{L}Q_{n,m}^*(x,y)\geq0$, *on* $[-1,0]^2$.

Furthermore it holds:

(1)

$$\left\|D_{1-}^{(\alpha_{1i},\alpha_{2j})}(f)-D_{1-}^{(\alpha_{1i},\alpha_{2j})}Q_{n,m}^*\right\|_{\infty,[-1,1]^2}\leq$$

$$\frac{\widetilde{C}2^{(i+j)-(\alpha_{1i}+\alpha_{2j})}}{\Gamma(i-\alpha_{1i}+1)\,\Gamma\left(j-\alpha_{2j}+1\right)n^{r-i}m^{p-j}}\cdot\omega_{s,q}\left(f^{(r,p)};\frac{1}{n},\frac{1}{m}\right), \tag{5.26}$$

where $\widetilde{C}$ *is a constant that depends only on* r,p,s,q; $(h_1+1,h_2+1)\leq(i,j)\leq(r,p)$, *or* $0\leq i\leq h_1$, $h_2+1\leq j\leq p$, *or* $h_1+1\leq i\leq r$, $0\leq j\leq h_2$,

(2)

$$\left\|D_{1-}^{(\alpha_{1i},\alpha_{2j})}(f)-D_{1-}^{(\alpha_{11},\alpha_{2j})}Q_{n,m}^*\right\|_{\infty,[-1,1]^2}\leq$$

$$\frac{c_{ij}}{n^{r-v_1}m^{p-v_2}}\cdot\omega_{s,q}\left(f^{(r,p)};\frac{1}{n},\frac{1}{m}\right), \tag{5.27}$$

for $(1,1)\leq(i,j)\leq(h_1,h_2)$, *where* $c_{ij}=\widetilde{C}A_{ij}$, *with*

$$A_{ij} :=$$

$$\left\{\left[\sum_{\tau=h_1}^{v_1}\sum_{\mu=h_2}^{v_2}\frac{l_{\tau\mu}2^{(\tau+\mu)-(\alpha_{1\tau}+\alpha_{2\mu})}}{\Gamma(\tau-a_{1\tau}+1)\Gamma(\mu-\alpha_{2\mu}+1)}\right]\left(\sum_{k=0}^{h_1-i}\frac{2^{h_1-\alpha_{1i}-k}}{k!\Gamma(h_1-\alpha_{1i}-k+1)}\right)\cdot\right. \tag{5.28}$$

$$\left.\left(\sum_{\lambda=0}^{h_2-j}\frac{2^{h_2-\alpha_{2j}-\lambda}}{\lambda!\Gamma(h_2-\alpha_{2j}-\lambda+1)}\right)+\frac{2^{(i+j)-(\alpha_{1i}+\alpha_{2j})}}{\Gamma(i-\alpha_{1i}+1)\Gamma(j-\alpha_{2j}+1)}\right\},$$

(3)

$$\left\|f-Q^*_{n,m}\right\|_{\infty,[-1,1]^2}\leq\frac{c_{00}}{n^{r-v_1}m^{p-v_2}}\cdot\omega_{s,q}\left(f^{(r,p)};\frac{1}{n},\frac{1}{m}\right), \tag{5.29}$$

where $c_{00} := \widetilde{C}A_{00}$, *with*

$$A_{00} := \frac{1}{h_1!h_2!}\left(\sum_{\tau=h_1}^{v_1}\sum_{\mu=h_2}^{v_2}l_{\tau\mu}\frac{2^{(\tau+\mu)-(\alpha_{1\tau}+\alpha_{2\mu})}}{\Gamma(\tau-a_{1\tau}+1)\Gamma(\mu-\alpha_{2\mu}+1)}\right)+1,$$

(4)

$$\left\|D^{(0,\alpha_{2j})}_{1-}(f)-D^{(0,\alpha_{2j})}_{1-}Q^*_{n,m}\right\|_{\infty,[-1,1]^2}\leq$$

$$\frac{c_{0j}}{n^{r-v_1}m^{p-v_2}}\cdot\omega_{s,q}\left(f^{(r,p)};\frac{1}{n},\frac{1}{m}\right), \tag{5.30}$$

where $c_{0j} = \widetilde{C}A_{0j}$, $j = 1, ..., h_2$, *with*

$$A_{0j} := \left[\frac{1}{h_1!}\left(\sum_{\tau=h_1}^{v_1}\sum_{\mu=h_2}^{v_2}l_{\tau\mu}\frac{2^{(\tau+\mu)-(\alpha_{1\tau}+\alpha_{2\mu})}}{\Gamma(\tau-a_{1\tau}+1)\Gamma(\mu-\alpha_{2\mu}+1)}\right)\right. \tag{5.31}$$

$$\left.\left(\sum_{\lambda=0}^{h_2-j}\frac{2^{h_2-\alpha_{2j}-\lambda}}{\lambda!\Gamma(h_2-\alpha_{2j}-\lambda+1)}\right)+\frac{2^{j-\alpha_{2j}}}{\Gamma(j-\alpha_{2j}+1)}\right],$$

and

(5)

$$\left\|D^{(\alpha_{1i},0)}_{1-}(f)-D^{(\alpha_{1i},0)}_{1-}Q^*_{n,m}\right\|_{\infty,[-1,1]^2}\leq$$

$$\frac{c_{i0}}{n^{r-v_1}m^{p-v_2}}\cdot\omega_{s,q}\left(f^{(r,p)};\frac{1}{n},\frac{1}{m}\right), \tag{5.32}$$

where $c_{i0} = \widetilde{C}A_{i0}$, $i = 1, ..., h_1$, *with*

$$A_{i0} := \left[\frac{1}{h_2!} \left(\sum_{\tau=h_1}^{v_1} \sum_{\mu=h_2}^{v_2} l_{\tau\mu} \frac{2^{(\tau+\mu)-(\alpha_{1\tau}+\alpha_{2\mu})}}{\Gamma(\tau - a_{1\tau} + 1)\Gamma(\mu - \alpha_{2\mu} + 1)} \right) \cdot \right. \tag{5.33}$$

$$\left. \left(\sum_{k=0}^{h_1-i} \frac{2^{h_1-\alpha_{1i}-k}}{k!\Gamma(h_1 - \alpha_{1i} - k + 1)} \right) + \frac{2^{i-\alpha_{1i}}}{\Gamma(i - \alpha_{1i} + 1)} \right].$$

In this chapter we establish right side bivariate abstract fractional calculus high order monotone (constrained) approximation theory by pseudo-polynomials of Caputo type, and then we apply our results to bivariate Prabhakar fractional Calculus, bivariate generalized non-singular fractional calculus, and bivariate parametrized Caputo-Fabrizio non-singular fractional calculus.

Next, we build the related necessary fractional calculi background.

5.2 Bivariate Right Side Fractional Calculi

Here we need to be very specific in preparation for our main results.

5.2.1 *Bivariate Right Abstract Fractional Calculus*

Let h_1, h_2, v_1, v_2, r, p be integers, $0 \leq h_1 \leq v_1 \leq r$, $0 \leq h_2 \leq v_2 \leq p$ and let $f \in C^{r,p}([-1,1]^2)$. Here $h_1 \leq i \leq v_1$, $h_2 \leq j \leq v_2$. Let $\alpha_{1i}, \alpha_{2j} \geq 0$, $\alpha_{1i}, \alpha_{2j} \notin \mathbb{Z}_+$: $\lceil \alpha_{1i} \rceil = i$, $\lceil \alpha_{2j} \rceil = j$, $i = 0, 1, ..., r$; $j = 0, 1, ..., p$, ($\lceil \cdot \rceil$ is the ceiling of number), $\alpha_{10} = 0$, $\alpha_{20} = 0$.

Consider also the integrable functions $k_{1i} := K_{\alpha_{1i}}$, $k_{2j} := K_{\alpha_{2j}} : [0, 2] \to \mathbb{R}_+$, $i = 0, 1, ..., r$; $j = 0, 1, ..., p$.

We consider the abstract right side Caputo type bivariate fractional partial derivative of orders $(\alpha_{1i}, \alpha_{2j})$

$$ {}^{k_{1i}}_{k_{2j}}D_{1-}^{(\alpha_{1i},\alpha_{2j})} f(x) := (-1)^{i+j} \int_{x_1}^{1} \int_{x_2}^{1} k_{1i}(t_1 - x_1) k_{2j}(t_2 - x_2) \frac{\partial^{i+j} f(t_1, t_2)}{\partial t_1^i \partial t_2^j} dt_1 dt_2, \tag{5.34}$$

$\forall\, x = (x_1, x_2) \in [-1, 1]^2$.

We set

$$ {}^{k_{10}}_{k_{20}}D_{1-}^{(0,0)} f(x) := f(x),$$

$$ {}^{k_{1i}}_{k_{2j}}D_{1-}^{(i,j)} f(x) := (-1)^{i+j} \frac{\partial^{i+j} f(x_1, x_2)}{\partial x_1^i \partial x_2^j}, \quad \forall\, x = (x_1, x_2) \in [-1, 1]^2. \tag{5.35}$$

We also set

$$
{}_{k_{2j}}^{k_{1i}}D_{1-}^{(i,\alpha_{2j})} f(x) := (-1)^j \int_{x_2}^{1} k_{2j}(t_2 - x_2) \frac{\partial^{i+j} f(x_1, t_2)}{\partial x_1^i \partial t_2^j} dt_2, \tag{5.36}
$$

$$
{}_{k_{2j}}^{k_{1i}}D_{1-}^{(\alpha_{1i},j)} f(x) := (-1)^i \int_{x_1}^{1} k_{1i}(t_1 - x_1) \frac{\partial^{i+j} f(t_1, x_2)}{\partial t_1^i \partial x_2^j} dt_1, \tag{5.37}
$$

and in particular we define:

$$
{}_{k_{2j}}^{k_{10}}D_{1-}^{(0,\alpha_{2j})} f(x) := (-1)^j \int_{x_2}^{1} k_{2j}(t_2 - x_2) \frac{\partial^j f(x_1, t_2)}{\partial t_2^j} dt_2, \tag{5.38}
$$

$$
{}_{k_{20}}^{k_{1i}}D_{1-}^{(\alpha_{1i},0)} f(x) := (-1)^i \int_{x_1}^{1} k_{1i}(t_1 - x_1) \frac{\partial^i f(t_1, x_2)}{\partial t_1^i} dt_1, \tag{5.39}
$$

$\forall\, x = (x_1, x_2) \in [-1, 1]^2$.

We will assume that

$$
\int_0^1 k_{ih_i}(z)\, dz \geq 1, \text{ when } h_i \neq 0, \tag{5.40}
$$

where $i = 1, 2$.

In [3], we got that $0 < \Gamma\left(h_1 - \alpha_{1h_1} + 1\right), \Gamma\left(h_2 - \alpha_{2h_2} + 1\right) \leq 1$, where Γ is the gamma function, and there it is

$$
k_{ih_i}(z) = \frac{z^{h_i - \alpha_{ih_i} - 1}}{\Gamma\left(h_i - \alpha_{ih_i}\right)}, \quad i = 1, 2; \ \forall\, z \in [0, 2],
$$

and (5.40) is fulfilled.

5.2.2 *About Right Bivariate Fractional Calculus*

We consider the Prabhakar function (also known as the three parameter Mittag-Leffler function), (see [10], [13, p. 97])

$$E_{\alpha,\beta}^{\gamma}(z)=\sum_{k=0}^{\infty}\frac{(\gamma)_k}{k!\Gamma(\alpha k+\beta)}z^k, \tag{5.41}$$

where $\alpha,\beta>0$, $\gamma\in\mathbb{R}$, $z\in\mathbb{R}$, and $(\gamma)_k=\gamma(\gamma+1)\dots(\gamma+k-1)$, it is $E_{\alpha,\beta}^{0}(z)=\frac{1}{\Gamma(\beta)}$.

Let $a,b,c,d\in\mathbb{R}$, $a<b$, $c<d$; $f\in C^{N_1,N_2}([a,b]\times[c,d])$, $\rho_i,\mu_i>0$; $\gamma_i,\omega_i\in\mathbb{R}$; $N_i=\lceil\mu_i\rceil$, $\mu_i\notin\mathbb{N}$; $i=1,2$.

We define the bivariate Prabhakar-Caputo right partial fractional derivative of orders (μ_1,μ_2) as follows ($x=(x_1,x_2)\in[a,b]\times[c,d]$):

$$\left({}^{C}D_{(\rho_1,\rho_2),(\mu_1,\mu_2),(\omega_1,\omega_2),(b-,d-)}^{(\gamma_1,\gamma_2)}f\right)(x)=$$

$$(-1)^{N_1+N_2}\int_{x_1}^{b}\int_{x_2}^{d}(t_1-x_1)^{N_1-\mu_1-1}(t_2-x_2)^{N_2-\mu_2-1} \tag{5.42}$$

$$E_{\rho_1,N_1-\mu_1}^{-\gamma_1}\left[\omega_1(t_1-x_1)^{\rho_1}\right]E_{\rho_2,N_2-\mu_2}^{-\gamma_2}\left[\omega_2(t_2-x_2)^{\rho_2}\right]\frac{\partial^{N_1+N_2}f(t_1,t_2)}{\partial t_1^{N_1}\partial t_2^{N_2}}dt_1dt_2,$$

with

$$\left({}^{C}D_{(\rho_1,\rho_2),(0,0),(\omega_1,\omega_2),(b-,d-)}^{(\gamma_1,\gamma_2)}f\right)(x):=f(x);$$

$$\left({}^{C}D_{(\rho_1,\rho_2),(N_1,N_2),(\omega_1,\omega_2),(b-,d-)}^{(\gamma_1,\gamma_2)}f\right)(x):=(-1)^{N_1+N_2}\frac{\partial^{N_1+N_2}f(x_1,x_2)}{\partial x_1^{N_1}\partial x_2^{N_2}}, \tag{5.43}$$

where $N_1,N_2\in\mathbb{N}$, etc.

For the related univariate theory see [5, 18].

5.2.3 *About Right Generalized Non-singular Fractional Calculus*

Here we use the multivariate analogue of generalized Mittag-Leffler function, see [19], defined for $\lambda,\gamma_j,\rho_j,z_j\in\mathbb{C}$, $Re(\rho_j)>0$ ($j=1,\dots,m$) in terms of a multiple series of the form:

$$E_{(\rho_j),\lambda}^{(\gamma_j)}(z_1,\dots,z_m)=E_{(\rho_1,\dots,\rho_m),\lambda}^{(\gamma_1,\dots,\gamma_m)}(z_1,\dots,z_m)=$$

$$\sum_{k_1,\dots,k_m=0}^{\infty} \frac{(\gamma_1)_{k_1} \dots (\gamma_m)_{k_m}}{\Gamma\left(\lambda + \sum_{j=1}^{m} k_j \rho_j\right)} \frac{z_1^{k_1} \dots z_m^{k_m}}{k_1! \dots k_m!}, \tag{5.44}$$

where $(\gamma_j)_{k_j}$ is the Pochhammer symbol. By [22, p. 157], (5.44) converges for $Re(\rho_j) > 0, j = 1, \dots, m$.

In particular we will use $\theta = 1, 2$; $E^{(\gamma_{\theta 1},\dots,\gamma_{\theta m})}_{(\rho_\theta,\dots,\rho_\theta),\lambda_\theta}\left[\omega_{\theta 1} t_\theta^{\rho_\theta}, \dots, \omega_{\theta m} t_\theta^{\rho_\theta}\right]$, denoted by $E^{(\gamma_{\theta j})}_{(\rho_\theta),\lambda_\theta}\left[\omega_{\theta 1} t_\theta^{\rho_\theta}, \dots, \omega_{\theta m} t_\theta^{\rho_\theta}\right]$, where $0 < \rho_\theta < 1, t_\theta \geq 0, \lambda_\theta > 0, \gamma_{\theta j} \in \mathbb{R}$ with $(\gamma_{\theta j})_{k_{\theta j}} := \gamma_{\theta j}(\gamma_{\theta j} + 1) \dots (\gamma_{\theta j} + k_{\theta j} - 1), \omega_{\theta j} \in \mathbb{R} - \{0\}$, for $j = 1, \dots, m$.

Let $f \in C^{N_1,N_2}([-1,1]^2)$, $0 < \mu_\theta \notin \mathbb{N}$ and $\lceil \mu_\theta \rceil = N_\theta \in \mathbb{N}$; $\theta = 1, 2$.

We define the bivariate Caputo type generalized right partial fractional derivative with non-singular kernel of order (μ_1, μ_2), as follows:

$$D^{(\mu_1,\mu_2)}_{1-} f(x) := {}^{(\gamma_{1j})(\omega_{1j})CA}_{(\gamma_{2j})(\omega_{2j})} D^{(\mu_1,\mu_2),(\lambda_1,\lambda_2)}_{1-} f(x) := \frac{A(1 - n_1 + \mu_1, 1 - N_2 + \mu_2)}{(N_1 - \mu_1)(N_2 - \mu_2)}$$

$$(-1)^{N_1+N_2} \int_{x_1}^{1} \int_{x_2}^{1} \prod_{\theta=1}^{2} E^{(\gamma_{\theta j})}_{(1-N_\theta+\mu_\theta),\lambda_\theta} \left[\frac{-\omega_{\theta 1}(1 - N_\theta + \mu_\theta)}{N_\theta - \mu_\theta}(t_\theta - x_\theta)^{1-N_\theta+\mu_\theta}\right.$$

$$\left., \dots, \frac{-\omega_{\theta m}(1 - N_\theta + \mu_\theta)}{N_\theta - \mu_\theta}(t_\theta - x_\theta)^{1-N_\theta+\mu_\theta}\right] \frac{\partial^{N_1+N_2} f(t_1, t_2)}{\partial t_1^{N_1} \partial t_2^{N_2}} dt_1 dt_2, \tag{5.45}$$

$\forall\, x = (x_1, x_2) \in [-1, 1]^2$, where $A := A(1 - n_1 + \mu_1, 1 - N_2 + \mu_2)$ is a normalizing constant. Without loss of generality we assume that $A > 0$.

We set $D^{(0,0)}_{1-} f = f$, $D^{(N_1,N_2)}_{1-} f = (-1)^{N_1+N_2} \frac{\partial^{N_1+N_2} f}{\partial x_1^{N_1} \partial x_2^{N_2}}$, when $N_1, N_2 \in \mathbb{N}$, etc.

For the univariate theory see the related [4, 6, 9].

5.2.4 *Bivariate Right Parametrized Caputo-Fabrizio Type Non-singular Kernel Left Partial Fractional Derivative of Orders* (μ_1, μ_2)

Let $f \in C^{N_1,N_2}([-1,1]^2)$, $\mathbb{N} \not\ni \mu_1, \mu_2 > 0$, $\lceil \mu_\theta \rceil = N_\theta \in \mathbb{N}$; $\omega_\theta < 0$; $\theta = 1, 2$.

It is given by

$${}^{CF}_{(\omega_1,\omega_2)} D^{(\mu_1,\mu_2)}_{1-} f(x) := \frac{(-1)^{N_1+N_2}}{(N_1 - \mu_1)(N_2 - \mu_2)}$$

$$\int_{x_1}^{1}\int_{x_2}^{1}\prod_{\theta=1}^{2}\left[\exp\left(-\frac{(1-N_\theta+\mu_\theta)\,\omega_\theta}{N_\theta-\mu_\theta}(t_\theta-x_\theta)\right)\right]\frac{\partial^{N_1+N_2}f(t_1,t_2)}{\partial t_1^{N_1}\partial t_2^{N_2}}dt_1dt_2, \quad (5.46)$$

$\forall\, x=(x_1,x_2)\in[-1,1]^2$, and
${}^{CF}_{(\omega_1,\omega_2)}D_{1-}^{(0,0)}f=f$, ${}^{CF}_{(\omega_1,\omega_2)}D_{1-}^{(N_1,N_2)}f=(-1)^{N_1+N_2}\frac{\partial^{N_1+N_2}f}{\partial x_1^{N_1}\partial x_2^{N_2}}$, etc.
For the univariate case see [6, 15].
We make

Remark 5.9 Right Fractional Calculi Sects. 5.2.2–5.2.4 are special cases of the right abstract fractional calculus Sect. 5.2.2. The abstract important condition (5.40) is fulfilled by: in Sect. 5.2.2 for large enough $\omega_\theta>0$, $\theta=1,2$; in Sect. 5.2.3 for small enough $\omega_{\theta j}<0$, $\theta=1,2$; $j=1,\ldots,m$; and in Sect. 5.2.4 for small enough $\omega_\theta<0$, $\theta=1,2$. For details see [6] and chap. 1.

5.3 Main Result

We present the following right bivariate abstract fractional monotone (constrained) approximation result.

Theorem 5.10 *Let h_1,h_2,v_1,v_2,r,p be integers, $0\le h_1\le v_1\le r$, $0\le h_2\le v_2\le p$ and let $f\in C^{r,p}([-1,1]^2)$, with $f^{(r,p)}$ having a mixed modulus of smoothness $\omega_{s,q}(f^{(r,p)};\delta_1,\delta_2)$, $\delta_1,\delta_2>0$ there, $s,q\in\mathbb{N}$. Let $\alpha_{ij}(x,y)$, $i=h_1,h_1+1,\ldots,v_1$; $j=h_2,h_2+1,\ldots,v_2$ be real-valued functions, defined and bounded in $[-1,1]^2$ and suppose $\alpha_{h_1h_2}$ is either $\ge\alpha>0$ or $\le\beta<0$ throughout $[-1,0]^2$. Assume that h_1, h_2 are even. Here $n,m\in\mathbb{N}:n\ge\max\{4(r+1),r+s\}$, $m\ge\max\{4(p+1),p+q\}$. Set*

$$l_{ij}:=\sup_{(x,y)\in[-1,1]^2}\left|\alpha_{h_1h_2}^{-1}(x,y)\,\alpha_{ij}(x,y)\right|<\infty \quad (5.47)$$

for all $h_1\le i\le v_1$, $h_2\le j\le v_2$. Let $0<\alpha_{1i},\alpha_{2j}\notin\mathbb{N}$ with $\lceil\alpha_{1i}\rceil=i$, $\lceil\alpha_{2j}\rceil=j$, $j=1,\ldots,r$; $j=1,\ldots,p$ ($\lceil\cdot\rceil$ is the ceiling of the number) and $\alpha_{10}=\alpha_{20}=0$.

Consider the abstract right fractional bivariate differential operator

$$L^*:=\sum_{i=h1}^{v_1}\sum_{j=h_2}^{v_2}\alpha_{ij}(x,y)\;{}^{k_{1i}}_{k_{2j}}D_{1-}^{(\alpha_{1i},\alpha_{2j})}. \quad (5.48)$$

Assume $L^*f(x,y)\ge0$, on $[-1,0]^2$. There exists a pseudo-polynomial of degree $\le(n,m)$

$$Q_{n,m}^*:=Q_{n,m}^*(f)\in(P_n\otimes C([-1,1])+C([-1,1])\otimes P_m)$$

such that $L^*Q_{n,m}^*(x,y)\ge0$, on $[-1,0]^2$.

We set

$$\lambda_{1i} := \int_0^2 k_{1i}(z)\,dz, \quad \lambda_{2j} := \int_0^2 k_{2j}(z)\,dz, \tag{5.49}$$

for $i = 1, ..., r;\ j = 1, ..., p.$

Set also $\lambda_{10} = \lambda_{20} = 1.$

Furthermore it holds:

(1) if $(h_1 + 1, h_2 + 1) \leq (i, j) \leq (r, p)$, *or* $0 \leq i \leq h_1$, $h_2 + 1 \leq j \leq p$, *or* $h_1 + 1 \leq i \leq r$, $0 \leq j \leq h_2$, *then*

$$\left\| {}_{k_{2j}}^{k_{1i}}D_{1-}^{(\alpha_{1i},\alpha_{2j})} f - {}_{k_{2j}}^{k_{1i}}D_{1-}^{(\alpha_{1i},\alpha_{2j})} Q_{n,m}^{*} \right\|_{\infty,[-1,1]^2} \leq$$

$$\lambda_{1i}\lambda_{2j} \frac{\widetilde{C}}{n^{r-i} m^{p-j}} \omega_{s,q}\left(f^{(r,p)}; \frac{1}{n}, \frac{1}{m}\right), \tag{5.50}$$

(2) if $(1, 1) \leq (i, j) \leq (h_1, h_2)$, *then*

$$\left\| {}_{k_{2j}}^{k_{1i}}D_{1-}^{(\alpha_{1i},\alpha_{2j})} f - {}_{k_{2j}}^{k_{1i}}D_{1-}^{(\alpha_{1i},\alpha_{2j})} Q_{n,m}^{*} \right\|_{\infty,[-1,1]^2} \leq \frac{\widetilde{C}\omega_{s,q}\left(f^{(r,p)}; \frac{1}{n}, \frac{1}{m}\right)}{n^{r-v_1} m^{p-v_2}}$$

$$\left\{ \left(\frac{\sum_{i_*=h_1}^{v_1} \sum_{j_*=h_2}^{v_2} l_{i_* j_*} \lambda_{1i_*} \lambda_{2j_*}}{h_1! h_2!} \right) \left\| {}_{k_{2j}}^{k_{1i}}D_{1-}^{(\alpha_{1i},\alpha_{2j})} \left(x^{h_1} y^{h_2}\right) \right\|_{\infty,[-1,1]^2} + \lambda_{1i}\lambda_{2j} \right\}, \tag{5.51}$$

(3) it holds

$$\left\| f - Q_{n,m}^{*} \right\|_{\infty,[-1,1]^2} \leq \frac{\widetilde{C}\omega_{s,q}\left(f^{(r,p)}; \frac{1}{n}, \frac{1}{m}\right)}{n^{r-v_1} m^{p-v_2}} \left[\frac{\sum_{i_*=h_1}^{v_1} \sum_{j_*=h_2}^{v_2} l_{i_* j_*} \lambda_{1i_*} \lambda_{2j_*}}{h_1! h_2!} + 1 \right], \tag{5.52}$$

(4) when $i = 0$, $j = 1, ..., h_2$, *we get*

$$\left\| {}_{k_{2j}}^{k_{10}}D_{1-}^{(0,\alpha_{2j})} f - {}_{k_{2j}}^{k_{10}}D_{1-}^{(0,\alpha_{2j})} Q_{n,m}^{*} \right\|_{\infty,[-1,1]^2} \leq \frac{\widetilde{C}\omega_{s,q}\left(f^{(r,p)}; \frac{1}{n}, \frac{1}{m}\right)}{n^{r-v_1} m^{p-v_2}}$$

$$\left[\left(\frac{\sum_{i_*=h_1}^{v_1} \sum_{j_*=h_2}^{v_2} l_{i_* j_*} \lambda_{1i_*} \lambda_{2j_*}}{h_1! h_2!} \right) \left\| {}_{k_{2j}}^{k_{10}}D_{1-}^{(0,\alpha_{2j})} y^{h_2} \right\|_{\infty,[-1,1]} + \lambda_{2j} \right], \tag{5.53}$$

and

(5) the case of $j=0$, $i=1,\dots,h_1$ *follows, it holds:*

$$\left\| {}_{k_{20}}^{k_{1i}}D_{1-}^{(\alpha_{1i},0)}f - {}_{k_{20}}^{k_{1i}}D_{1-}^{(\alpha_{1i},0)}Q_{n,m}^{*}\right\|_{\infty,[-1,1]^2} \leq \frac{\widetilde{C}\omega_{s,q}\left(f^{(r,p)};\frac{1}{n},\frac{1}{m}\right)}{n^{r-v_1}m^{p-v_2}}$$

$$\left[\left(\frac{\sum_{i_*=h_1}^{v_1}\sum_{j_*=h_2}^{v_2} l_{i_*j_*}\lambda_{1i_*}\lambda_{2j_*}}{h_1!h_2!}\right)\left\| {}_{k_{20}}^{k_{1i}}D_{1-}^{(\alpha_{1i},0)}x^{h_1}\right\|_{\infty,[-1,1]} + \lambda_{1i}\right]. \tag{5.54}$$

Proof By Corollary 5.3 there exists

$$Q_{n,m} \equiv Q_{n,m}(f) \in (P_n \otimes C([-1,1]) + C([-1,1]) \otimes P_m)$$

such that

$$\left\|f^{(i,j)} - Q_{n,m}^{(i,j)}\right\|_{\infty} \leq \frac{\widetilde{C}}{n^{r-i}m^{p-j}} \cdot \omega_{s,q}\left(f^{(r,p)};\frac{1}{n},\frac{1}{m}\right), \tag{5.55}$$

for all $(0,0) \leq (i,j) \leq (r,p)$, while $Q_{n,m} \in C^{r,p}\left([-1,1]^2\right)$. Here $\widetilde{C}$ depends only on r,p,s,q, where $n \geq \max\{4(r+1), r+s\}$ and $m \geq \max\{4(p+1), p+q\}$, with $r,p \in \mathbb{Z}_+$, $s,q \in \mathbb{N}$, $f \in C^{r,p}\left([-1,1]^2\right)$.

Indeed by [12] we have that $Q_{n,m}^{(r,p)}$ is continuous on $[-1,1]^2$.

We observe the following ($i=1,\dots,r$; $j=1,\dots,p$)

$$\left| {}_{k_{2j}}^{k_{1i}}D_{1-}^{(\alpha_{1i},\alpha_{2j})}f(x_1,x_2) - {}_{k_{2j}}^{k_{1i}}D_{1-}^{(\alpha_{1i},\alpha_{2j})}Q_{n,m}(x_1,x_2)\right| =$$

$$\left|(-1)^{i+j}\int_{x_1}^{1}\int_{x_2}^{1} k_{1i}(t_1-x_1)k_{2j}(t_2-x_2)\frac{\partial^{i+j}f(t_1,t_2)}{\partial t_1^i\partial t_2^j}dt_1dt_2 - \right.$$

$$\left.(-1)^{i+j}\int_{x_1}^{1}\int_{x_2}^{1} k_{1i}(t_1-x_1)k_{2j}(t_2-x_2)\frac{\partial^{i+j}Q_{n,m}(t_1,t_2)}{\partial t_1^i\partial t_2^j}dt_1dt_2\right| = \tag{5.56}$$

$$\left|\int_{x_1}^{1}\int_{x_2}^{1} k_{1i}(t_1-x_1)k_{2j}(t_2-x_2)\left(\frac{\partial^{i+j}f(t_1,t_2)}{\partial t_1^i\partial t_2^j} - \frac{\partial^{i+j}Q_{n,m}(t_1,t_2)}{\partial t_1^i\partial t_2^j}\right)dt_1dt_2\right| \leq$$

$$\int_{x_1}^{1}\int_{x_2}^{1} k_{1i}(t_1-x_1)k_{2j}(t_2-x_2)\left|\frac{\partial^{i+j}f(t_1,t_2)}{\partial t_1^i\partial t_2^j} - \frac{\partial^{i+j}Q_{n,m}(t_1,t_2)}{\partial t_1^i\partial t_2^j}\right|dt_1dt_2 \leq$$

$$\left(\int_{x_1}^{1}\int_{x_2}^{1} k_{1i}(t_1-x_1)\,k_{2j}(t_2-x_2)\,dt_1dt_2\right)\frac{\widetilde{C}}{n^{r-i}m^{p-j}}\omega_{s,q}\left(f^{(r,p)};\frac{1}{n},\frac{1}{m}\right)=$$

$$\left(\int_{x_1}^{1} k_{1i}(t_1-x_1)\,dt_1\right)\left(\int_{x_2}^{1} k_{2j}(t_2-x_2)\,dt_2\right)\frac{\widetilde{C}}{n^{r-i}m^{p-j}}\omega_{s,q}\left(f^{(r,p)};\frac{1}{n},\frac{1}{m}\right)= \tag{5.57}$$

$$\left(\int_{0}^{1-x_1} k_{1i}(z)\,dz\right)\left(\int_{0}^{1-x_2} k_{2j}(z)\,dz\right)\frac{\widetilde{C}}{n^{r-i}m^{p-j}}\omega_{s,q}\left(f^{(r,p)};\frac{1}{n},\frac{1}{m}\right)\leq$$

$$\left(\int_{0}^{2} k_{1i}(z)\,dz\right)\left(\int_{0}^{2} k_{2j}(z)\,dz\right)\frac{\widetilde{C}}{n^{r-i}m^{p-j}}\omega_{s,q}\left(f^{(r,p)};\frac{1}{n},\frac{1}{m}\right).$$

We have proved that

$$\left|{}_{k_{2j}}^{k_{1i}}D_{1-}^{(\alpha_{1i},\alpha_{2j})}f(x_1,x_2)-{}_{k_{2j}}^{k_{1i}}D_{1-}^{(\alpha_{1i},\alpha_{2j})}Q_{n,m}(x_1,x_2)\right|\leq \tag{5.58}$$

$$\left(\int_{0}^{2} k_{1i}(z)\,dz\right)\left(\int_{0}^{2} k_{2j}(z)\,dz\right)\frac{\widetilde{C}}{n^{r-i}m^{p-j}}\omega_{s,q}\left(f^{(r,p)};\frac{1}{n},\frac{1}{m}\right),$$

$\forall\,(x_1,x_2)\in[-1,1]^2$; $i=1,\dots,r$; $j=1,\dots,p$.

So we have proved that there exists $Q_{n,m}$ such that

$$\left\|{}_{k_{2j}}^{k_{1i}}D_{1-}^{(\alpha_{1i},\alpha_{2j})}(f)-{}_{k_{2j}}^{k_{1i}}D_{1-}^{(\alpha_{1i},\alpha_{2j})}(Q_{n,m})\right\|_{\infty,[-1,1]^2}\leq \tag{5.59}$$

$$\left(\int_{0}^{2} k_{1i}(z)\,dz\right)\left(\int_{0}^{2} k_{2j}(z)\,dz\right)\frac{\widetilde{C}}{n^{r-i}m^{p-j}}\omega_{s,q}\left(f^{(r,p)};\frac{1}{n},\frac{1}{m}\right),$$

$i=1,\dots,r$; $j=1,\dots,p$.

We call

$$\lambda_{1i}:=\int_{0}^{2} k_{1i}(z)\,dz,\quad \lambda_{2j}:=\int_{0}^{2} k_{2j}(z)\,dz, \tag{5.60}$$

for $i=1,\dots,r$; $j=1,\dots,p$, as in (5.49).

We also set $\lambda_{10}=\lambda_{20}=1$.

Thus the following inequality is valid in general:

$$\left\| {}^{k_{1i}}_{k_{2j}}D^{(\alpha_{1i},\alpha_{2j})}_{1-}(f) - {}^{k_{1i}}_{k_{2j}}D^{(\alpha_{1i},\alpha_{2j})}_{1-}(Q_{n,m}) \right\|_{\infty,[-1,1]^2} \leq \tag{5.61}$$

$$\lambda_{1i}\lambda_{2j}\frac{\widetilde{C}}{n^{r-i}m^{p-j}}\omega_{s,q}\left(f^{(r,p)};\frac{1}{n},\frac{1}{m}\right),$$

for $i = 0, 1, ..., r$; $j = 0, 1, ..., p$.

Define

$$\rho_{n,m} := \widetilde{C}\omega_{s,q}\left(f^{(r,p)};\frac{1}{n},\frac{1}{m}\right)\left[\sum_{i=h_1}^{v_1}\sum_{j=h_2}^{v_2}\left(l_{ij}\lambda_{1i}\lambda_{2j}n^{i-r}m^{j-p}\right)\right]. \tag{5.62}$$

(I) Suppose, throughout $[-1, 0]^2$, $\alpha_{h_1h_2}(x, y) \geq \alpha > 0$. Let $Q^*_{n,m}(x, y) \in (P_n \otimes C([-1, 1]) + C([-1, 1]) \otimes P_m)$, $(x, y) \in [-1, 1]^2$, as in (5.61), so that

$$\left\| {}^{k_{1i}}_{k_{2j}}D^{(\alpha_{1i},\alpha_{2j})}_{1-}\left(f(x, y) + \rho_{n,m}\frac{x^{h_1}}{h_1!}\frac{y^{h_2}}{h_2!}\right) - {}^{k_{1i}}_{k_{2j}}D^{(\alpha_{1i},\alpha_{2j})}_{1-}Q^*_{n,m}(x, y) \right\|_{\infty,[-1,1]^2} \leq$$

$$\lambda_{1i}\lambda_{2j}\frac{\widetilde{C}}{n^{r-i}m^{p-j}}\omega_{s,q}\left(f^{(r,p)};\frac{1}{n},\frac{1}{m}\right) =: T_{ij}, \tag{5.63}$$

for $i = 0, 1, ..., r$; $j = 0, 1, ..., p$.

If $(h_1 + 1, h_2 + 1) \leq (i, j) \leq (r, p)$, or $0 \leq i \leq h_1$, $h_2 + 1 \leq j \leq p$, or $h_1 + 1 \leq i \leq r$, $0 \leq j \leq h_2$ we get from the last

$$\left\| {}^{k_{1i}}_{k_{2j}}D^{(\alpha_{1i},\alpha_{2j})}_{1-}(f) - {}^{k_{1i}}_{k_{2j}}D^{(\alpha_{1i},\alpha_{2j})}_{1-}Q^*_{n,m} \right\|_{\infty,[-1,1]^2} \leq$$

$$\lambda_{1i}\lambda_{2j}\frac{\widetilde{C}}{n^{r-i}m^{p-j}}\omega_{s,q}\left(f^{(r,p)};\frac{1}{n},\frac{1}{m}\right), \tag{5.64}$$

proving (5.50).

If $(0, 0) \leq (i, j) \leq (h_1, h_2)$, we get that

$$\left\| {}^{k_{1i}}_{k_{2j}}D^{(\alpha_{1i},\alpha_{2j})}_{1-}f + \frac{\rho_{n,m}}{h_1!h_2!}\,{}^{k_{1i}}_{k_{2j}}D^{(\alpha_{1i},\alpha_{2j})}_{1-}\left(x^{h_1}y^{h_2}\right) - {}^{k_{1i}}_{k_{2j}}D^{(\alpha_{1i},\alpha_{2j})}_{1-}Q^*_{n,m}(x, y) \right\|_{\infty,[-1,1]^2}$$

$$\leq T_{ij}. \tag{5.65}$$

That is for $(1, 1) \leq (i, j) \leq (h_1, h_2)$, we have

$$\left\| {}^{k_{1i}}_{k_{2j}}D^{(\alpha_{1i},\alpha_{2j})}_{1-}f - {}^{k_{1i}}_{k_{2j}}D^{(\alpha_{1i},\alpha_{2j})}_{1-}Q^*_{n,m} \right\|_{\infty,[-1,1]^2} \leq$$

$$\frac{\rho_{n,m}}{h_1!h_2!}\left\| {}_{k_{2j}}^{k_{1i}}D_{1-}^{(\alpha_{1i},\alpha_{2j})}\left(x^{h_1}y^{h_2}\right)\right\|_{\infty,[-1,1]^2}+T_{ij}=$$

$$\frac{\widetilde{C}\omega_{s,q}\left(f^{(r,p)};\frac{1}{n},\frac{1}{m}\right)}{h_1!h_2!}\left[\sum_{i_*=h_1}^{v_1}\sum_{j_*=h_2}^{v_2}\left(l_{i_*j_*}\lambda_{1i_*}\lambda_{2j_*}n^{i_*-r}m^{j_*-p}\right)\right]$$

$$\left\| {}_{k_{2j}}^{k_{1i}}D_{1-}^{(\alpha_{1i},\alpha_{2j})}\left(x^{h_1}y^{h_2}\right)\right\|_{\infty,[-1,1]^2}+\lambda_{1i}\lambda_{2j}\frac{\widetilde{C}}{n^{r-i}m^{p-j}}\omega_{s,q}\left(f^{(r,p)};\frac{1}{n},\frac{1}{m}\right)= \tag{5.66}$$

$$\widetilde{C}\omega_{s,q}\left(f^{(r,p)};\frac{1}{n},\frac{1}{m}\right)\left\{\frac{1}{h_1!h_2!}\left[\sum_{i_*=h_1}^{v_1}\sum_{j_*=h_2}^{v_2}l_{i_*j_*}\lambda_{1i_*}\lambda_{2j_*}n^{i_*-r}m^{j_*-p}\right]\right.$$

$$\left.\left\| {}_{k_{2j}}^{k_{1i}}D_{1-}^{(\alpha_{1i},\alpha_{2j})}\left(x^{h_1}y^{h_2}\right)\right\|_{\infty,[-1,1]^2}+\frac{\lambda_{1i}\lambda_{2j}}{n^{r-i}m^{p-j}}\right\}\leq$$

$$\frac{\widetilde{C}\omega_{s,q}\left(f^{(r,p)};\frac{1}{n},\frac{1}{m}\right)}{n^{r-v_1}m^{p-v_2}}\left\{\left(\frac{\sum_{i_*=h_1}^{v_1}\sum_{j_*=h_2}^{v_2}l_{i_*j_*}\lambda_{1i_*}\lambda_{2j_*}}{h_1!h_2!}\right)\right. \tag{5.67}$$

$$\left.\left\| {}_{k_{2j}}^{k_{1i}}D_{1-}^{(\alpha_{1i},\alpha_{2j})}\left(x^{h_1}y^{h_2}\right)\right\|_{\infty,[-1,1]^2}+\lambda_{1i}\lambda_{2j}\right\},$$

proving (5.51).

If $i=j=0$, from (5.63) we obtain

$$\left\|f+\frac{\rho_{n,m}}{h_1!h_2!}x^{h_1}y^{h_2}-Q_{n,m}^*\right\|_{\infty,[-1,1]^2}\leq\frac{\widetilde{C}}{n^rm^p}\omega_{s,q}\left(f^{(r,p)};\frac{1}{n},\frac{1}{m}\right), \tag{5.68}$$

and

$$\left\|f-Q_{n,m}^*\right\|_{\infty,[-1,1]^2}\leq\frac{\rho_{n,m}}{h_1!h_2!}+\frac{\widetilde{C}}{n^rm^p}\omega_{s,q}\left(f^{(r,p)};\frac{1}{n},\frac{1}{m}\right)=$$

$$\frac{\widetilde{C}\omega_{s,q}\left(f^{(r,p)};\frac{1}{n},\frac{1}{m}\right)}{h_1!h_2!}\left[\sum_{i_*=h_1}^{v_1}\sum_{j_*=h_2}^{v_2}l_{i_*j_*}\lambda_{1i_*}\lambda_{2j_*}n^{i_*-r}m^{j_*-p}\right]$$

$$+\frac{\widetilde{C}}{n^rm^p}\omega_{s,q}\left(f^{(r,p)};\frac{1}{n},\frac{1}{m}\right)=$$

$$\widetilde{C}\omega_{s,q}\left(f^{(r,p)};\frac{1}{n},\frac{1}{m}\right)\left[\frac{1}{h_1!h_2!}\sum_{i_*=h_1}^{v_1}\sum_{j_*=h_2}^{v_2} l_{i_*j_*}\lambda_{1i_*}\lambda_{2j_*}n^{i_*-r}m^{j_*-p}+\frac{1}{n^r m^p}\right]\leq \tag{5.69}$$

$$\frac{\widetilde{C}\omega_{s,q}\left(f^{(r,p)};\frac{1}{n},\frac{1}{m}\right)}{n^{r-v_1}m^{p-v_2}}\left[\frac{\sum_{i_*=h_1}^{v_1}\sum_{j_*=h_2}^{v_2} l_{i_*j_*}\lambda_{1i_*}\lambda_{2j_*}}{h_1!h_2!}+1\right],$$

proving (5.52).

Next case is for $i=0, j=1, ..., h_2$, from (5.63) we get

$$\left\| {}_{k_{2j}}^{k_{10}}D_{1-}^{(0,\alpha_{2j})}f+\frac{\rho_{n,m}x^{h_1}}{h_1!h_2!}\,{}_{k_{2j}}^{k_{10}}D_{1-}^{(0,\alpha_{2j})}y^{h_2}-{}_{k_{2j}}^{k_{10}}D_{1-}^{(0,\alpha_{2j})}Q_{n,m}^{*}\right\|_{\infty,[-1,1]^2}\leq \tag{5.70}$$

$$\lambda_{2j}\frac{\widetilde{C}}{n^r m^{p-j}}\omega_{s,q}\left(f^{(r,p)};\frac{1}{n},\frac{1}{m}\right).$$

Therefore we have that

$$\left\| {}_{k_{2j}}^{k_{10}}D_{1-}^{(0,\alpha_{2j})}f-{}_{k_{2j}}^{k_{10}}D_{1-}^{(0,\alpha_{2j})}Q_{n,m}^{*}\right\|_{\infty,[-1,1]^2}\leq$$

$$\frac{\rho_{n,m}}{h_1!h_2!}\left\| {}_{k_{2j}}^{k_{10}}D_{1-}^{(0,\alpha_{2j})}y^{h_2}\right\|_{\infty,[-1,1]}+\lambda_{2j}\frac{\widetilde{C}}{n^r m^{p-j}}\omega_{s,q}\left(f^{(r,p)};\frac{1}{n},\frac{1}{m}\right)=$$

$$\frac{\widetilde{C}\omega_{s,q}\left(f^{(r,p)};\frac{1}{n},\frac{1}{m}\right)}{h_1!h_2!}\left[\sum_{i_*=h_1}^{v_1}\sum_{j_*=h_2}^{v_2}\left(l_{i_*j_*}\lambda_{1i_*}\lambda_{2j_*}n^{i_*-r}m^{j_*-p}\right)\right]$$

$$\left\| {}_{k_{2j}}^{k_{10}}D_{1-}^{(0,\alpha_{2j})}y^{h_2}\right\|_{\infty,[-1,1]}+\lambda_{2j}\frac{\widetilde{C}}{n^r m^{p-j}}\omega_{s,q}\left(f^{(r,p)};\frac{1}{n},\frac{1}{m}\right)=$$

$$\widetilde{C}\omega_{s,q}\left(f^{(r,p)};\frac{1}{n},\frac{1}{m}\right)\left[\frac{\left[\sum_{i_*=h_1}^{v_1}\sum_{j_*=h_2}^{v_2}\left(l_{i_*j_*}\lambda_{1i_*}\lambda_{2j_*}n^{i_*-r}m^{j_*-p}\right)\right]}{h_1!h_2!}\right.$$

$$\left.\left\| {}_{k_{2j}}^{k_{10}}D_{1-}^{(0,\alpha_{2j})}y^{h_2}\right\|_{\infty,[-1,1]}+\frac{\lambda_{2j}}{n^r m^{p-j}}\right]\leq \tag{5.71}$$

$$\frac{\widetilde{C}\omega_{s,q}\left(f^{(r,p)};\frac{1}{n},\frac{1}{m}\right)}{n^{r-v_1}m^{p-v_2}}\left[\frac{\left(\sum_{i_*=h_1}^{v_1}\sum_{j_*=h_2}^{v_2} l_{i_*j_*}\lambda_{1i_*}\lambda_{2j_*}\right)}{h_1!h_2!}\right.$$

$$\left.\left\|{}_{k_{2j}}^{k_{10}}D_{1-}^{(0,\alpha_{2j})}y^{h_2}\right\|_{\infty,[-1,1]}+\lambda_{2j}\right],$$

proving (5.53).

The case of $j=0$, $i=1,...,h_1$, is met similarly as in (5.71). Namely we get

$$\left\|{}_{k_{20}}^{k_{1i}}D_{1-}^{(\alpha_{1i},0)}f-{}_{k_{20}}^{k_{1i}}D_{1-}^{(\alpha_{1i},0)}Q_{n,m}^{*}\right\|_{\infty,[-1,1]^2}\leq$$

$$\frac{\widetilde{C}\omega_{s,q}\left(f^{(r,p)};\frac{1}{n},\frac{1}{m}\right)}{n^{r-v_1}m^{p-v_2}}\left[\left(\frac{\sum_{i_*=h_1}^{v_1}\sum_{j_*=h_2}^{v_2} l_{i_*j_*}\lambda_{1i_*}\lambda_{2j_*}}{h_1!h_2!}\right)\right. \tag{5.72}$$

$$\left.\left\|{}_{k_{20}}^{k_{1i}}D_{1-}^{(\alpha_{1i},0)}x^{h_1}\right\|_{\infty,[-1,1]}+\lambda_{1i}\right],$$

proving (5.54).

So if $(x,y)\in[-1,0]^2$, we can write

$$\alpha_{h_1h_2}^{-1}(x,y)\,L^*\left(Q_{n,m}^{*}(x,y)\right)=\alpha_{h_1h_2}^{-1}(x,y)\,L^*\left(f(x,y)\right)+$$

$$\frac{\rho_{n,m}}{h_1!h_2!}\,{}_{k_{2h_2}}^{k_{1h_1}}D_{1-}^{(\alpha_{1h_1},\alpha_{2h_2})}\left(x^{h_1}y^{h_2}\right)+\sum_{i=h_1}^{v_1}\sum_{j=h_2}^{v_2}\alpha_{h_1h_2}^{-1}(x,y)\,\alpha_{ij}(x,y)\cdot$$

$$\left[{}_{k_{2j}}^{k_{1i}}D_{1-}^{(\alpha_{1i},\alpha_{2j})}Q_{n,m}^{*}(x,y)-{}_{k_{2j}}^{k_{1i}}D_{1-}^{(\alpha_{1i},\alpha_{2j})}f(x,y)-\frac{\rho_{n,m}}{h_1!h_2!}\,{}_{k_{2j}}^{k_{1i}}D_{1-}^{(\alpha_{1i},\alpha_{2j})}\left(x^{h_1}y^{h_2}\right)\right] \tag{5.73}$$

(by $L^*f\geq 0$)

$$\overset{(5.63)}{\geq}\frac{\rho_{n,m}}{h_1!h_2!}\,{}_{k_{2h_2}}^{k_{1h_1}}D_{1-}^{(\alpha_{1h_1},\alpha_{2h_2})}\left(x^{h_1}y^{h_2}\right)-$$

$$\left(\sum_{i=h_1}^{v_1}\sum_{j=h_2}^{v_2} l_{ij}\lambda_{1i}\lambda_{2j}n^{i-r}m^{j-p}\right)\widetilde{C}\omega_{s,q}\left(f^{(r,p)};\frac{1}{n},\frac{1}{m}\right)=$$

$$\rho_{n,m}\left(\frac{{}_{k_{2h_2}}^{k_{1h_1}}D_{1-}^{(\alpha_{1h_1},\alpha_{2h_2})}\left(x^{h_1}y^{h_2}\right)}{h_1!h_2!}-1\right)= \tag{5.74}$$

$$\rho_{n,m}\left[\frac{\left({}_{k_{20}}^{k_{1h_1}}D_{1-}^{(\alpha_{1h_1},0)}x^{h_1}\right)\left({}_{k_{2h_2}}^{k_{10}}D_{1-}^{(0,\alpha_{2h_2})}y^{h_2}\right)}{h_1!h_2!}-1\right]=:\varphi.$$

If $h_1=h_2=0$, then $\alpha_{1h_1}=\alpha_{2h_2}=0$, and $\varphi=0$.

If $h_1=0$, and $h_2\neq 0$, then

$$\varphi=\rho_{n,m}\left[\frac{{}_{k_{2h_2}}^{k_{10}}D_{1-}^{(0,\alpha_{2h_2})}y^{h_2}}{h_2!}-1\right]=$$

$$\rho_{n,m}\left[(-1)^{h_2}\int_y^1 k_{2h_2}(t-y)\,dt-1\right]\overset{(h_2\text{ is even})}{=}\rho_{n,m}\left(\int_0^{1-y}k_{2h_2}(z)\,dz-1\right)\geq \tag{5.75}$$

$$\rho_{n,m}\left(\int_0^1 k_{2h_2}(z)\,dz-1\right)\geq 0,$$

by the assumption (5.40).

Similarly, we treat the case $h_1\neq 0$, $h_2=0$.

When $h_1,h_2\neq 0$, then we have

$$\varphi\overset{(h_1,h_2\text{ are even})}{=}\rho_{n,m}\left[\left(\int_x^1 k_{1h_1}(t-x)\,dt\right)\left(\int_y^1 k_{2h_2}(t-y)\,dt\right)-1\right]= \tag{5.76}$$

$$\rho_{n,m}\left[\left(\int_0^{1-x}k_{1h_1}(z)\,dz\right)\left(\int_0^{1-y}k_{2h_2}(z)\,dz\right)-1\right]\geq$$

$$\rho_{n,m}\left[\left(\int_0^1 k_{1h_1}(z)\,dz\right)\left(\int_0^1 k_{2h_2}(z)\,dz\right)-1\right]\geq 0,$$

by the assumption (5.40).

So in all four cases we get that

$$L^*\left(Q_{n,m}^*(x,y)\right)\geq 0,\ \forall\ (x,y)\in[-1,0]^2. \tag{5.77}$$

(II) Suppose, throughout $[-1,0]^2$, $\alpha_{h_1h_2}(x,y) \leq \beta < 0$. Let $Q^*_{n,m}(x,y) \in (P_n \otimes C([-1,1]) + C([-1,1]) \otimes P_m)$, $(x,y) \in [-1,1]^2$, as in (5.61), so that

$$\left\| {}_{k_{2j}}^{k_{1i}}D_{1-}^{(\alpha_{1i},\alpha_{2j})}\left(f(x,y) - \rho_{n,m}\frac{x^{h_1}}{h_1!}\frac{y^{h_2}}{h_2!}\right) - {}_{k_{2j}}^{k_{1i}}D_{1-}^{(\alpha_{1i},\alpha_{2j})}Q^{**}_{n,m}(x,y)\right\|_{\infty,[-1,1]^2} \leq$$

$$\lambda_{1i}\lambda_{2j}\frac{\widetilde{C}}{n^{r-i}m^{p-j}}\omega_{s,q}\left(f^{(r,p)};\frac{1}{n},\frac{1}{m}\right), \tag{5.78}$$

for $i = 0, 1, ..., r$, $j = 0, 1, ..., p$.

Similarly, we get as in the first case $\geq \alpha > 0$, the inequalities of simultaneous fractional convergence, see {(5.51), (5.67)}, {(5.52), (5.69)}, {(5.53), (5.71)}, {(5.54), (5.72)}.

So if $(x,y) \in [-1,0]^2$ we can write

$$\alpha^{-1}_{h_1h_2}(x,y)L^*\left(Q^*_{n,m}(x,y)\right) = \alpha^{-1}_{h_1h_2}(x,y)L^*(f(x,y)) -$$

$$\frac{\rho_{n,m}}{h_1!h_2!}{}_{k_{2h_2}}^{k_{1h_1}}D_{1-}^{(\alpha_{1h_1},\alpha_{2h_2})}\left(x^{h_1}y^{h_2}\right) + \sum_{i=h_1}^{v_1}\sum_{j=h_2}^{v_2}\alpha^{-1}_{h_1h_2}(x,y)\alpha_{ij}(x,y)\cdot \tag{5.79}$$

$$\left[{}_{k_{2j}}^{k_{1i}}D_{1-}^{(\alpha_{1i},\alpha_{2j})}Q^*_{n,m}(x,y) - {}_{k_{2j}}^{k_{1i}}D_{1-}^{(\alpha_{1i},\alpha_{2j})}f(x,y) + \frac{\rho_{n,m}}{h_1!h_2!}{}_{k_{2j}}^{k_{1i}}D_{1-}^{(\alpha_{1i},\alpha_{2j})}\left(x^{h_1}y^{h_2}\right)\right]$$

(by $L^*f \geq 0$)

$$\overset{(5.78)}{\leq} -\frac{\rho_{n,m}}{h_1!h_2!}{}_{k_{2h_2}}^{k_{1h_1}}D_{1-}^{(\alpha_{1h_1},\alpha_{2h_2})}\left(x^{h_1}y^{h_2}\right) +$$

$$\left(\sum_{i=h_1}^{v_1}\sum_{j=h_2}^{v_2}l_{ij}\lambda_{1i}\lambda_{2j}n^{r-i}m^{p-j}\right)\widetilde{C}\omega_{s,q}\left(f^{(r,p)};\frac{1}{n},\frac{1}{m}\right) = \tag{5.80}$$

$$\rho_{n,m}\left[1 - \frac{{}_{k_{2h_2}}^{k_{1h_1}}D_{1-}^{(\alpha_{1h_1},\alpha_{2h_2})}\left(x^{h_1}y^{h_2}\right)}{h_1!h_2!}\right] =$$

$$\rho_{n,m}\left[1 - \frac{\left({}_{k_{20}}^{k_{1h_1}}D_{1-}^{(\alpha_{1h_1},0)}x^{h_1}\right)\left({}_{k_{2h_2}}^{k_{10}}D_{1-}^{(0,\alpha_{2h_2})}y^{h_2}\right)}{h_1!h_2!}\right] =: \psi.$$

If $h_1 = h_2 = 0$, then $\alpha_{1h_1} = \alpha_{2h_2} = 0$, and $\psi = 0$.

If $h_1 = 0$, and $h_2 \neq 0$, then

$$\psi = \rho_{n,m}\left[1 - \frac{\frac{k_{10}}{k_{2h_2}} D_{1-}^{(0,\alpha_{2h_2})} y^{h_2}}{h_2!}\right] = \rho_{n,m}\left[1 - \int_y^1 k_{2h_2}(t-y)\,dt\right] = \tag{5.81}$$

$$\rho_{n,m}\left[1 - \int_0^{1-y} k_{2h_2}(z)\,dz\right] \leq \rho_{n,m}\left[1 - \int_0^1 k_{2h_2}(z)\,dz\right] \leq 0,$$

see (5.40).

Similarly, we treat the case $h_1 \neq 0$, $h_2 = 0$.

When $h_1, h_2 \neq 0$, then we have

$$\psi = \rho_{n,m}\left[1 - \left(\int_x^1 k_{1h_1}(t-x)\,dt\right)\left(\int_y^1 k_{2h_2}(t-y)\,dt\right)\right] =$$

$$\rho_{n,m}\left[1 - \left(\int_0^{1-x} k_{1h_1}(z)\,dz\right)\left(\int_0^{1-y} k_{2h_2}(z)\,dz\right)\right] \leq \tag{5.82}$$

$$\rho_{n,m}\left[1 - \left(\int_0^1 k_{1h_1}(z)\,dz\right)\left(\int_0^1 k_{2h_2}(z)\,dz\right)\right] \leq 0,$$

by the assumption (5.40).

So in all four cases we proved again that

$$L^*\left(Q_{n,m}^*(x,y)\right) \geq 0, \ \forall\ (x,y) \in [-1,0]^2. \tag{5.83}$$

The proof is complete. ■

Conclusion 5.11 Clearly Theorem 5.10 generalizes Theorem 5.8 to many right side fractional calculi, opening new avenues of fractional research activity. The approximating pseudo-polynomial $Q_{n,m}^*$ depends on f, $\rho_{n,m}$, h_1, h_2; which $\rho_{n,m}$ depends on $\widetilde{C}$ (which depends on r, p, s, q), $f, n, m, l_{ij}, \lambda_{1i}, \lambda_{2j}$; and which: λ_{1i} depends on k_{1i} and λ_{2j} depends on k_{2j}. That is $Q_{n,m}^*$ depends on the type of bivariate right side fractional calculus we use.

Consequently, Theorem 5.10 is valid at least for the following important bivariate right fractional linear differential operators:

(1)

$$L_1^* := \sum_{i=h_1}^{\nu_1} \sum_{j=h_2}^{\nu_2} \alpha_{ij}(x, y) \left[{}^{C}D^{(\gamma_1,\gamma_2)}_{(\rho_1,\rho_2),(\alpha_{1i},\alpha_{2j}),(\omega_1,\omega_2),(1-,1-)} \right], \tag{5.84}$$

where $\rho_1, \rho_2 > 0$, $\gamma_1, \gamma_2 < 0$, and $\omega_1, \omega_2 > 0$ large enough (from right bivariate Prabhakar fractional calculus, see (5.42));

(2)

$$L_2^* := \sum_{i=h_1}^{\nu_1} \sum_{j=h_2}^{\nu_2} \alpha_{ij}(x, y) \left[D_{1-}^{(\alpha_{1i},\alpha_{2j})} \right], \tag{5.85}$$

(see (5.45)) where $\theta = 1, 2$; $\gamma_{\theta j} > 0$, $j = 1, \ldots, m$; $\lambda_\theta = 1$, $0 < \rho_\theta < 1$; and small enough $\omega_{\theta j} < 0, j = 1, \ldots, m$ (from right bivariate generalized non-singular fractional calculus);

and

(3)

$$L_3^* := \sum_{i=h_1}^{\nu_1} \sum_{j=h_2}^{\nu_2} \alpha_{ij}(x, y) \left[{}^{CF}_{(\omega_1,\omega_2)}D_{1-}^{(\alpha_{1i},\alpha_{2j})} \right], \tag{5.86}$$

(see (5.46)) for small enough $\omega_1, \omega_2 < 0$ (from right bivariate parametrized Caputo-Fabrizio non-singular kernel fractional calculus).

Our developed right bivariate abstract fractional monotone approximation theory by pseudopolynomials with its applications, involves weaker conditions than the one with ordinary partial derivatives, see Theorem 5.4, and can manage many diverse general cases in a multitude of complex settings and environments.

References

1. Anastassiou, G.A.: Monotone approximation by pseudopolynomials. In: Approximation Theory, pp. 5–11. Academic Press, New York (1991)
2. Anastassiou, G.A.: Bivariate monotone approximation. Proc. Amer. Math. Soc. **112**(4), 959–964 (1991)
3. Anastassiou, G.A.: Bivariate right fractional pseudo-polynomial monotone approximation. In: Intelligent Mathematics II: Applied Mathematics and Approximation Theory, pp. 15–31. AMAT, Ankara, May 2015, Springer, New York (2016)
4. Anastassiou, G.A.: Multiparameter fractional differentiation with non singular kernel. Issues. Anal. (2021) (accepted)
5. Anastassiou, G.A.: Foundations of generalized Prabhakar-Hilfer fractional calculus with applications. Cubo (2021) (accepted)
6. Anastassiou, G.A.: Univariate simultaneous high order abstract fractional monotone approximation with applications (2021) (submitted)
7. Anastassiou, G.A.: Abstract bivariate right fractional pseudo-polynomial monotone constrained approximation and applications (2021) (submitted)

8. Anastassiou, G.A., Shisha, O.: Monotone approximation with linear differential operators. J. Approx. Theory **44**, 391–393 (1985)
9. Atangana, A., Baleanu, D.: New fractional derivatives with nonlocal and non-singular kernel: theory and application to heat transfer model. Therm. Sci. **20**(2), 763–769 (2016)
10. Giusti, A., et al.: A practical Guide to Prabhakar fractional calculus. Fract. Calculus App. Anal. **23**(1), 9–54 (2020)
11. Gonska, H.H.: Simultaneously approximation by algebraic blending functions. In: Alfred Haar memorial conference, Budapest, vol 49, pp. 363–382. Coloquia Mathematica Soc. Janos Bolyai. North-Holand, Amsterdam (1985)
12. Gonska, H.H.: Personal communication with author (2014)
13. Gorenflo, R., Kilbas, A., Mainardi, F., Rogosin, S.: Mittag-Leffler Functions, Related Topics and Applications. Springer, Heidelberg, New York (2014)
14. Iqbal, S., Krulic, K., Pecaric, J.: On an inequality of H.G. Hardy. J. Inequal. Appl. **2010**(Article ID 264347), 23 p
15. Losada,J., Nieto, J.J.: Properties of a new fractional derivative without singular kernel. Progr. Fract. Differ. Appl. **1**(2), 87–92 (2015)
16. Marchaud, A.: Differences et deerivees d'une fonction de deux variables. C.R. Acad. Sci. **178** 1467–1470 (1924)
17. Marchaud, A.: Sur les derivees et sur les differences des fonctions de variables reelles. J. Math. Pures Appl. **6**, 337–425 (1927)
18. Polito, F., Tomovski, Z.: Some properties of Prabhakar-type fractional calculus operators. Fract. Differ. Calculus **6**(1), 73–94 (2016)
19. Saxena, R.K., Kalla, S.L.: Ravi Saxena. Multivariate analogue of generalized Mittag-Leffler function. Integr. Transf. Spec. Funct. **22**(7), 533–548 (2011)
20. Schumaker, L.L.: Spline Functions: Basic Theory. Wiley, New York (1981)
21. Shisha, O.: Monotone approximation. Pac. J. Math. **15**, 667–671 (1965)
22. Srivastava, H.M., Daoust, M.C.: A note on the convergence of Kompe' de Feriet's double hypergeometric series. Math. Nachr. **53**, 151–159 (1972)

Chapter 6
Bivariate Polynomial Abstract Left and Right Fractional Monotone Constrained Approximation

Let $f \in C^{r,p}\left([0,1]^2\right)$, $r, p \in \mathbb{N}$, and let L^* be an abstract linear left or right fractional mixed partial differential operator such that $L^*(f) \geq 0$, for all (x, y) in a critical region of $[0,1]^2$ that depends on L^*. Then there exists a sequentce of two-dimensional polynomials $Q_{\overline{m_1},\overline{m_2}}(x, y)$ with $L^*\left(Q_{\overline{m_1},\overline{m_2}}(x, y)\right) \geq 0$ there, where $\overline{m_1}, \overline{m_2} \in \mathbb{N}$ such that $\overline{m_1} > r$, $\overline{m_2} > p$, so that f is approximated left or right abstract fractionally simultaneously and uniformly by $Q_{\overline{m_1},\overline{m_2}}$ on $[0,1]^2$. This restricted left or right abstract fractional approximation is achieved quantitatively by the use of a suitable integer partial derivatives two-dimensional first modulus of continuity. This monotone constrained fractional approximation applies to a wide range of Caputo type fractional calculi of singular or non-singular kernels. It follows [4].

6.1 Introduction

The topic of monotone approximation started in [8] has become a major trend in approximation theory. A typical problem in this subject is: given a positive integer k, approximate a given function whose kth derivative is ≥ 0 by polynomials having this property.

In [5] the authors replaced the kth derivative with a linear differential operator of order k. We mention this motivating result.

Theorem 6.1 *Let h, k, p be integers, $0 \leq h \leq k \leq p$ and let f be a real function, $f^{(p)}$ continuous in $[-1,1]$ with first modulus of continuity $\omega_1\left(f^{(p)}, x\right)$ there. Let $a_j(x)$, $j = h, h+1, \ldots, k$ be real functions, defined and bounded on $[-1,1]$ and assume $a_h(x)$ is either $\geq$ some number $\alpha > 0$ or $\leq$ some number $\beta < 0$ throughout $[-1,1]$. Consider the operator*

G. A. Anastassiou, *Abstract Fractional Monotone Approximation, Theory and Applications*, Studies in Systems, Decision and Control 411,
https://doi.org/10.1007/978-3-030-95943-2_6

$$L=\sum_{j=h}^{k}a_j(x)\left[\frac{d^j}{dx^j}\right]$$

and suppose, throughout $[-1,1]$,

$$L(f)\geq 0. \tag{6.1}$$

Then, for every integer $n\geq 1$, *there is a real polynomial* $Q_n(x)$ *of degree* $\leq n$ *such that*

$$L(Q_n)\geq 0 \text{ throughout } [-1,1]$$

and

$$\max_{-1\leq x\leq 1}|f(x)-Q_n(x)|\leq Cn^{k-p}\omega_1\left(f^{(p)},\frac{1}{n}\right),$$

where C is independent of n or f.

We need

Definition 6.2 (*Stancu* [9]) Let $f\in C\left([0,1]^2\right)$, $[0,1]^2=[0,1]\times[0,1]$, where (x_1,y_1), $(x_2,y_2)\in[0,1]^2$ and $\delta_1,\delta_2\geq 0$. The first modulus of continuity of f is defined as follows:

$$\omega_1(f,\delta_1,\delta_2)=\sup_{\substack{|x_1-x_2|\leq\delta_1\\|y_1-y_2|\leq\delta_2}}|f(x_1,y_1)-f(x_2,y_2)|.$$

Definition 6.3 Let f be a real-valued function defined on $[0,1]^2$ and let m,n be two positive integers. Let $B_{m,n}$ be the Bernstein (polynomial) operator of order (m,n) given by

$$B_{m,n}(f;x,y)=\sum_{i=0}^{m}\sum_{j=0}^{n}f\left(\frac{i}{m},\frac{j}{n}\right)\cdot\binom{m}{i}\cdot\binom{n}{j}\cdot x^i(1-x)^{m-i}y^j(1-y)^{n-j}. \tag{6.2}$$

For integers $r,s\geq 0$, we denote by $f^{(r,s)}$ the differential operator of order (r,s) given by

$$f^{(r,s)}(x,y)=\frac{\partial^{r+s}f(x,y)}{\partial x^r\partial y^s}.$$

We use

Theorem 6.4 (Badea and Badea [6]) *It holds that*

$$\left\|f^{(k,l)}-\left(B_{m,n}f\right)^{(k,l)}\right\|_\infty\leq t(k,l)\cdot$$

$$\omega_1\left(f^{(k,l)};\frac{1}{\sqrt{m-k}},\frac{1}{\sqrt{n-l}}\right)+\max\left\{\frac{k(k-1)}{m},\frac{l(l-1)}{n}\right\}\cdot\left\|f^{(k,l)}\right\|_\infty, \tag{6.3}$$

where $m>k\geq 0$, $n>l\geq 0$ are integers, f is a real-valued function on $[0,1]^2$ such that $f^{(k,l)}$ is continuous, and t is a positive real-valued function on $\mathbb{Z}_+^2=\{0,1,2,...\}^2$. Here $\|\cdot\|_\infty$ is the supremum norm on $[0,1]^2$.

Denote $C^{r,p}\left([0,1]^2\right):=\{f:[0,1]^2\to\mathbb{R};f^{(k,l)}$ is continuous for $0\leq k\leq r, 0\leq l\leq p\}$.

In [1] the author proved the following motivational result.

Theorem 6.5 *Let h_1,h_2,v_1,v_2,r,p be integers, $0\leq h_1\leq v_1\leq r$, $0\leq h_2\leq v_2\leq p$ and let $f\in C^{r,p}\left([0,1]^2\right)$. Let $\alpha_{i,j}(x,y)$, $i=h_1,h_1+1,...,v_1$; $j=h_2,h_2+1,...,v_2$ be real-valued functions, defined and bounded in $[0,1]^2$ and suppose $\alpha_{h_1h_2}$ is either $\geq\alpha>0$ or $\leq\beta<0$ throughout $[0,1]^2$. Consider the operator*

$$L=\sum_{i=h_1}^{v_1}\sum_{j=h_2}^{v_2}\alpha_{ij}(x,y)\frac{\partial^{i+j}}{\partial x^i\partial y^j} \tag{6.4}$$

and assume that throughout $[0,1]^2$,

$$L(f)\geq 0.$$

Then for integers m,n with $m>r$, $n>p$, there exists a polynomial $Q_{m,n}(x,y)$ of degree (m,n) such that $L\left(Q_{m,n}(x,y)\right)\geq 0$ throughout $[0,1]^2$ and

$$\left\|f^{(k,l)}-Q_{m,n}^{(k,l)}\right\|_\infty\leq\frac{P_{m,n}(L,f)}{(h_1-k)!\,(h_2-l)!}+M_{m,n}^{k,l}(f), \tag{6.5}$$

all $(0,0)\leq(k,l)\leq(h_1,h_2)$. Furthermore we get

$$\left\|f^{(k,l)}-Q_{m,n}^{(k,l)}\right\|_\infty\leq M_{m,n}^{k,l}(f), \tag{6.6}$$

for all $(h_1+1,h_2+1)\leq(k,l)\leq(r,p)$. Also (6.6) is true whenever $0\leq k\leq h_1$, $h_2+1\leq l\leq p$ or $h_1+1\leq k\leq r$, $0\leq l\leq h_2$. Here

$$M_{m,n}^{k,l}\equiv M_{m,n}^{k,l}(f)\equiv t(k,l)\cdot$$

$$\omega_1\left(f^{(k,l)};\frac{1}{\sqrt{m-k}},\frac{1}{\sqrt{n-l}}\right)+\max\left\{\frac{k(k-1)}{m},\frac{l(l-1)}{n}\right\}\cdot\left\|f^{(k,l)}\right\|_\infty, \tag{6.7}$$

and

$$P_{m,n}\equiv P_{m,n}(L,f)\equiv\sum_{i=h_1}^{v_1}\sum_{j=h_2}^{v_2}l_{ij}\cdot M_{m,n}^{i,j}, \tag{6.8}$$

where t is a positive real-valued function on $\mathbb{Z}_+^2$ *and*

$$l_{ij} \equiv \sup_{(x,y)\in[0,1]^2} \left|\alpha_{h_1 h_2}^{-1}(x,y)\cdot\alpha_{ij}(x,y)\right| < \infty. \tag{6.9}$$

In [2] we extended Theorem 6.5 to the fractional level. Indeed there L is replaced by L^*, a linear left Caputo fractional mixed partial differential operator. Now the monotonicity property is only true on a critical region of $[0,1]^2$ that depends on L^* parameters. Simultaneous fractional convergence remains true on all of $[0,1]^2$.

We need

Definition 6.6 Let $\alpha_1, \alpha_2 > 0$; $\alpha = (\alpha_1, \alpha_2)$, $f \in C\left([0,1]^2\right)$, and let $x = (x_1, x_2)$, $(t_1, t_2) \in [0,1]^2$. We define the left mixed Riemann-Liouville fractional two dimensional integral of order α (see also [7]):

$$\left(I_{0+}^{\alpha} f\right)(x) := \frac{1}{\Gamma(\alpha_1)\Gamma(\alpha_2)} \int_0^{x_1}\int_0^{x_2} (x_1 - t_1)^{\alpha_1 - 1}(x_2 - t_2)^{\alpha_2 - 1} f(t_1, t_2)\, dt_1 dt_2, \tag{6.10}$$

with $x_1, x_2 > 0$.

Notice here that $I_{0+}^{\alpha}(|f|) < \infty$.

Definition 6.7 ([2]) Let $\alpha_1, \alpha_2 > 0$ with $\lceil\alpha_1\rceil = m_1$, $\lceil\alpha_2\rceil = m_2$, ($\lceil\cdot\rceil$ ceiling of the number). Let here $f \in C^{m_1,m_2}\left([0,1]^2\right)$. We consider the left (Caputo type) fractional partial derivative:

$$D_{*0}^{(\alpha_1,\alpha_2)} f(x) := \frac{1}{\Gamma(m_1 - \alpha_1)\Gamma(m_2 - \alpha_2)} \cdot$$

$$\int_0^{x_1}\int_0^{x_2} (x_1 - t_1)^{m_1 - \alpha_1 - 1}(x_2 - t_2)^{m_2 - \alpha_2 - 1} \frac{\partial^{m_1+m_2} f(t_1, t_2)}{\partial t_1^{m_1}\partial t_2^{m_2}} dt_1 dt_2, \tag{6.11}$$

$\forall\, x = (x_1, x_2) \in [0,1]^2$, where Γ is the gamma function

$$\Gamma(\nu) = \int_0^{\infty} e^{-t} t^{\nu-1} dt, \quad \nu > 0. \tag{6.12}$$

We set

$$D_{*0}^{(0,0)} f(x) := f(x), \quad \forall\, x \in [0,1]^2; \tag{6.13}$$

$$D_{*0}^{(m_1,m_2)} f(x) := \frac{\partial^{m_1+m_2} f(x)}{\partial x_1^{m_1}\partial x_2^{m_2}}, \quad \forall\, x \in [0,1]^2. \tag{6.14}$$

Definition 6.8 ([2]) We also set

$$D_{*0}^{(0,\alpha_2)} f(x) := \frac{1}{\Gamma(m_2-\alpha_2)} \int_0^{x_2} (x_2-t_2)^{m_2-\alpha_2-1} \frac{\partial^{m_2} f(x_1,t_2)}{\partial t_2^{m_2}} dt_2, \tag{6.15}$$

$$D_{*0}^{(\alpha_1,0)} f(x) := \frac{1}{\Gamma(m_1-\alpha_1)} \int_0^{x_1} (x_1-t_1)^{m_1-\alpha_1-1} \frac{\partial^{m_1} f(t_1,x_2)}{\partial t_1^{m_1}} dt_1, \tag{6.16}$$

and

$$D_{*0}^{(m_1,\alpha_2)} f(x) := \frac{1}{\Gamma(m_2-\alpha_2)} \int_0^{x_2} (x_2-t_2)^{m_2-\alpha_2-1} \frac{\partial^{m_1+m_2} f(x_1,t_2)}{\partial x_1^{m_1} \partial t_2^{m_2}} dt_2, \tag{6.17}$$

$$D_{*0}^{(\alpha_1,m_2)} f(x) := \frac{1}{\Gamma(m_1-\alpha_1)} \int_0^{x_1} (x_1-t_1)^{m_1-\alpha_1-1} \frac{\partial^{m_1+m_2} f(t_1,x_2)}{\partial t_1^{m_1} \partial x_2^{m_2}} dt_1. \tag{6.18}$$

The following result is the main motivation for this chapter:

Theorem 6.9 ([2]) *Let h_1, h_2, v_1, v_2, r, p be integers, $0 \le h_1 \le v_1 \le r$, $0 \le h_2 \le v_2 \le p$ and let $f \in C^{r,p}\left([0,1]^2\right)$. Let $\alpha_{ij}(x,y)$, $i = h_1, h_1+1, \ldots, v_1$; $j = h_2, h_2+1, \ldots, v_2$ be real valued functions, defined and bounded in $[0,1]^2$ and assume $\alpha_{h_1 h_2}$ is either $\ge \alpha > 0$ or $\le \beta < 0$ throughout $[0,1]^2$. Let $0 \le \alpha_{1h_1} \le h_1 < \alpha_{11} \le h_1+1 < \alpha_{12} \le h_1+2 < \alpha_{13} \le h_1+3 < \ldots < h_{1v_1} \le v_1 < \ldots < \alpha_{1r} \le r$, with $\lceil \alpha_{1h_1} \rceil = h_1$; $0 \le \alpha_{2h_2} \le h_2 < \alpha_{21} \le h_2+1 < \alpha_{22} \le h_2+2 < \alpha_{23} \le h_2+3 < \ldots < \alpha_{2v_2} \le v_2 < \ldots < \alpha_{2p} \le p$, with $\lceil \alpha_{2h_2} \rceil = h_2$. Consider the left fractional differential bivariate operator*

$$L^* := \sum_{i=h_1}^{v_1} \sum_{j=h_2}^{v_2} \alpha_{ij}(x,y) D_{*0}^{(\alpha_{1i},\alpha_{2j})}. \tag{6.19}$$

Let integers $\overline{m_1}, \overline{m_2}$ with $\overline{m_1} > r$, $\overline{m_2} > p$. Set

$$l_{ij} := \sup_{(x,y)\in[0,1]^2} \left|\alpha_{h_1 h_2}^{-1}(x,y) \cdot \alpha_{ij}(x,y)\right| < \infty.$$

Also set ($\lceil \alpha_{1i} \rceil = i$, $\lceil \alpha_{2j} \rceil = j$, $\lceil \cdot \rceil$ ceiling of number)

$$M_{\overline{m_1},\overline{m_2}}^{i,j} := M_{\overline{m_1},\overline{m_2}}^{i,j}(f) := \tag{6.20}$$

$$\frac{1}{\Gamma(i-\alpha_{1i}+1)\,\Gamma\left(j-\alpha_{2j}+1\right)} \left\{ t(i,j)\, \omega_1\left(f^{(i,j)}; \frac{1}{\sqrt{\overline{m_1}-i}}, \frac{1}{\sqrt{\overline{m_2}-j}}\right)\right.$$

$$+\max\left\{\frac{i(i-1)}{\overline{m_1}},\frac{j(j-1)}{\overline{m_2}}\right\}\cdot\left\|f^{(i,j)}\right\|_\infty\right\},$$

$i=h_1,...,v_1$; $j=h_2,...,v_2$.

Here t is a positive real-valued function on $\mathbb{Z}_+^2$, $\|\cdot\|_\infty$ is the supremum norm on $[0,1]^2$. Call

$$P_{\overline{m_1},\overline{m_2}}:=P_{\overline{m_1},\overline{m_2}}(f)=\sum_{i=h_1}^{v_1}\sum_{j=h_2}^{v_2}l_{ij}\cdot M_{\overline{m_1},\overline{m_2}}^{i,j}. \tag{6.21}$$

Then there exists a polynomial $Q_{\overline{m_1},\overline{m_2}}(x,y)$ of degree $(\overline{m_1},\overline{m_2})$ on $[0,1]^2$ such that

$$\left\|D_{*0}^{(\alpha_{1k},\alpha_{2l})}(f)-D_{*0}^{(\alpha_{1k},\alpha_{2l})}\left(Q_{\overline{m_1},\overline{m_2}}\right)\right\|_\infty\le$$

$$\frac{\Gamma(h_1-k+1)\,\Gamma(h_2-l+1)\,P_{\overline{m_1},\overline{m_2}}}{\Gamma(h_1-\alpha_{1k}+1)\,\Gamma(h_2-\alpha_{2l}+1)\,(h_1-k)!\,(h_2-l)!}+M_{\overline{m_1},\overline{m_2}}^{k,l}, \tag{6.22}$$

for $(0,0)\le(k,l)\le(h_1,h_2)$.

If $(h_1+1,h_2+1)\le(k,l)\le(r,p)$, or $0\le k\le h_1$, $h_2+1\le l\le p$, or $h_1+1\le k\le r$, $0\le l\le h_2$, we get

$$\left\|D_{*0}^{(\alpha_{1k},\alpha_{2l})}(f)-D_{*0}^{(\alpha_{1k},\alpha_{2l})}\left(Q_{\overline{m_1},\overline{m_2}}\right)\right\|_\infty\le M_{\overline{m_1},\overline{m_2}}^{k,l}. \tag{6.23}$$

By assuming $L^*(f(1,1))\ge0$, we get $L^*\left(Q_{\overline{m_1},\overline{m_2}}(1,1)\right)\ge0$.

Let $1\ge x,y>0$, with $\alpha_{1h_1}\ne h_1$ and $\alpha_{2h_2}\ne h_2$, such that

$$x\ge\left(\Gamma\left(h_1-\alpha_{1h_1}+1\right)\right)^{\frac{1}{\left(h_1-\alpha_{1h_1}\right)}}, \tag{6.24}$$

$$y\ge\left(\Gamma\left(h_2-\alpha_{2h_2}+1\right)\right)^{\frac{1}{\left(h_2-\alpha_{2h_2}\right)}},$$

and

$$L^*(f(x,y))\ge0.$$

Then

$$L^*\left(Q_{\overline{m_1},\overline{m_2}}(x,y)\right)\ge0.$$

Some notation follows:

Definition 6.10 Let f be a real-valued function defined on $[0,1]^2$ and let $\overline{m_1},\overline{m_2}\in\mathbb{N}$. Let $B_{\overline{m_1},\overline{m_2}}$ be the Bernstein (polynomial) operator of order $(\overline{m_1},\overline{m_2})$ given by

$$B_{\overline{m_1},\overline{m_2}}(f;x_1,x_2):=$$

$$\sum_{i_1=0}^{\overline{m_1}} \sum_{i_2=0}^{\overline{m_2}} f\left(\frac{i_1}{\overline{m_1}}, \frac{i_2}{\overline{m_2}}\right) \binom{\overline{m_1}}{i_1} \binom{\overline{m_2}}{i_2} x_1^{i_1} (1-x_1)^{\overline{m_1}-i_1} x_2^{i_2} (1-x_2)^{\overline{m_2}-i_2}. \quad (6.25)$$

In this chapter we generalize Theorem 6.9 to abstract kernels that can be singular or non-singular, again bivariate constrained monotonicity takes place over a critical region of $[0, 1]^2$, however bivariate abstract fractional simultaneous approximation is true over the whole of $[0, 1]^2$. We cover both the left and right sides of this bivariate fractional approximation.

We need the following abstract fractional background.

6.2 About Bivariate Abstract Fractional Calculus

Let h_1, h_2, v_1, v_2, r, p be integers, $0 \le h_1 \le v_1 \le r$, $0 \le h_2 \le v_2 \le p$ and let $f \in C^{r,p}\left([0, 1]^2\right)$. Here $h_1 \le i \le v_1, h_2 \le j \le v_2$. Let $\alpha_{1i}, \alpha_{2j} \ge 0, \alpha_{1i}, \alpha_{2j} \notin \mathbb{Z}_+ : \lceil \alpha_{1i} \rceil = i$, $\lceil \alpha_{2j} \rceil = j$, $i = 0, 1, ..., r$; $j = 0, 1, ..., p$, ($\lceil \cdot \rceil$ is the ceiling of number), $\alpha_{10} = 0$, $\alpha_{20} = 0$.

Consider also the integrable functions $k_{1i} := K_{\alpha_{1i}}, k_{2j} := K_{\alpha_{2j}} : [0, 1] \to \mathbb{R}_+, i = 0, 1, ..., r; j = 0, 1, ..., p$.

(I) We first consider the abstract left Caputo type bivariate fractional partial derivative of orders $\left(\alpha_{1i}, \alpha_{2j}\right)$

$$_{k_{2j}}^{k_{1i}}D_{*0}^{(\alpha_{1i},\alpha_{2j})} f(x) := \int_0^{x_1} \int_0^{x_2} k_{1i}(x_1 - t_1) k_{2j}(x_2 - t_2) \frac{\partial^{i+j} f(t_1, t_2)}{\partial t_1^i \partial t_2^j} dt_1 dt_2, \quad (6.26)$$

$\forall x = (x_1, x_2) \in [0, 1]^2$.

We set

$$_{k_{20}}^{k_{10}}D_{*0}^{(0,0)} f(x) := f(x),$$

$$_{k_{2j}}^{k_{1i}}D_{*0}^{(i,j)} f(x) := \frac{\partial^{i+j} f(x_1, x_2)}{\partial x_1^i \partial x_2^j}, \ \forall x = (x_1, x_2) \in [0, 1]^2. \quad (6.27)$$

We also set

$$_{k_{2j}}^{k_{1i}}D_{*0}^{(i,\alpha_{2j})} f(x) := \int_0^{x_2} k_{2j}(x_2 - t_2) \frac{\partial^{i+j} f(x_1, t_2)}{\partial x_1^i \partial t_2^j} dt_2, \quad (6.28)$$

$$_{k_{2j}}^{k_{1i}}D_{*0}^{(\alpha_{1i},j)} f(x) := \int_0^{x_1} k_{1i}(x_1 - t_1) \frac{\partial^{i+j} f(t_1, x_2)}{\partial t_1^i \partial x_2^j} dt_1, \quad (6.29)$$

and in particular we define:

$$ {}^{k_{10}}_{k_{2j}}D^{(0,\alpha_{2j})}_{*0} f(x) := \int_0^{x_2} k_{2j}(x_2 - t_2) \frac{\partial^j f(x_1, t_2)}{\partial t_2^j} dt_2, \quad (6.30) $$

$$ {}^{k_{1i}}_{k_{20}}D^{(\alpha_{1i},0)}_{*0} f(x) := \int_0^{x_1} k_{1i}(x_1 - t_1) \frac{\partial^i f(t_1, x_2)}{\partial t_1^i} dt_1, \quad (6.31) $$

$\forall\, x = (x_1, x_2) \in [0, 1]^2$.

We will assume that there exists a critical region $\varnothing \neq \Phi \subseteq [0, 1]^2$ such that

$$ \Phi := \left\{ (x, y) \in [0, 1]^2 : \int_0^x k_{1h_1}(z)\, dz, \ \int_0^y k_{2h_2}(z)\, dz \geq 1, \text{ for any of } h_1, h_2 \neq 0 \right\}. \quad (6.32) $$

In [2], we got that $0 < \Gamma\left(h_1 - \alpha_{1h_1} + 1\right), \Gamma\left(h_2 - \alpha_{2h_2} + 1\right) \leq 1$, where Γ is the gamma function, and there it is

$$ k_{ih_i}(z) = \frac{z^{h_i - \alpha_{ih_i} - 1}}{\Gamma\left(h_i - \alpha_{ih_i}\right)}, \quad i = 1, 2; \ \forall\, z \in [0, 1], \quad (6.33) $$

and it holds

$$ \int_0^1 k_{ih_i}(z)\, dz = \frac{1}{\Gamma\left(h_i - \alpha_{ih_i} + 1\right)} \geq 1, \quad (6.34) $$

proving there that $\Phi \neq \varnothing$, by $x = y = 1$.

Also in [2], see Theorem 6.9, when $\alpha_{1h_1} \neq h_1$ and $\alpha_{2h_2} \neq h_2$ and $0 < x, y < 1$ the critical region Φ there contains all (x, y) such that

$$ \left\{ \begin{array}{l} 1 > x \geq \left(\Gamma\left(h_1 - \alpha_{1h_1} + 1\right)\right)^{\frac{1}{\left(h_1 - \alpha_{1h_1}\right)}}, \\ 1 > y \geq \left(\Gamma\left(h_2 - \alpha_{2h_2} + 1\right)\right)^{\frac{1}{\left(h_2 - \alpha_{2h_2}\right)}}. \end{array} \right\}, \quad (6.35) $$

so again $\Phi \neq \varnothing$, non-trivially.

(II) We also consider the abstract right side Caputo type bivariate fractional partial derivative of orders $\left(\alpha_{1i}, \alpha_{2j}\right)$

$$ {}^{k_{1i}}_{k_{2j}}D^{(\alpha_{1i},\alpha_{2j})}_{1-} f(x) := (-1)^{i+j} \int_{x_1}^1 \int_{x_2}^1 k_{1i}(t_1 - x_1)\, k_{2j}(t_2 - x_2) \frac{\partial^{i+j} f(t_1, t_2)}{\partial t_1^i \partial t_2^j} dt_1 dt_2, \quad (6.36) $$

$\forall\, x = (x_1, x_2) \in [0, 1]^2$.

We set

$$
\begin{gathered}
{}^{k_{10}}_{k_{20}}D^{(0,0)}_{1-}f(x) := f(x), \\
{}^{k_{1i}}_{k_{2j}}D^{(i,j)}_{1-}f(x) := (-1)^{i+j}\frac{\partial^{i+j}f(x_1,x_2)}{\partial x_1^i \partial x_2^j}, \quad \forall\, x = (x_1,x_2) \in [0,1]^2 .
\end{gathered}
\tag{6.37}
$$

We also set

$$
{}^{k_{1i}}_{k_{2j}}D^{(i,\alpha_{2j})}_{1-}f(x) := (-1)^j \int_{x_2}^{1} k_{2j}(t_2 - x_2)\frac{\partial^{i+j}f(x_1,t_2)}{\partial x_1^i \partial t_2^j}dt_2, \tag{6.38}
$$

$$
{}^{k_{1i}}_{k_{2j}}D^{(\alpha_{1i},j)}_{1-}f(x) := (-1)^i \int_{x_1}^{1} k_{1i}(t_1 - x_1)\frac{\partial^{i+j}f(t_1,x_2)}{\partial t_1^i \partial x_2^j}dt_1, \tag{6.39}
$$

and in particular we define:

$$
{}^{k_{10}}_{k_{2j}}D^{(0,\alpha_{2j})}_{1-}f(x) := (-1)^j \int_{x_2}^{1} k_{2j}(t_2 - x_2)\frac{\partial^{j}f(x_1,t_2)}{\partial t_2^j}dt_2, \tag{6.40}
$$

$$
{}^{k_{1i}}_{k_{20}}D^{(\alpha_{1i},0)}_{1-}f(x) := (-1)^i \int_{x_1}^{1} k_{1i}(t_1 - x_1)\frac{\partial^{i}f(t_1,x_2)}{\partial t_1^i}dt_1, \tag{6.41}
$$

$\forall\, x = (x_1,x_2) \in [0,1]^2$.

We will assume that there exists a critical region $\varnothing \neq \Psi \subseteq [0,1]^2$ such that

$$
\Psi := \left\{ (x,y) \in [0,1]^2 : \int_0^{1-x} k_{1h_1}(z)\,dz,\ \int_0^{1-y} k_{2h_2}(z)\,dz \geq 1, \text{ for any of } h_1, h_2 \neq 0 \right\}. \tag{6.42}
$$

When $x = y = 0$ we have the conditions in (6.42) fulfilled by (6.33), (6.34), so there $\Psi \neq \varnothing$, see [3].

Also in [3], when $\alpha_{1h_1} \neq h_1$ and $\alpha_{2h_2} \neq h_2$ and $0 < x, y < 1$ the critical region Ψ there contains all (x,y) such that

$$
\left\{
\begin{array}{l}
1 - x \geq \left(\Gamma\left(h_1 - \alpha_{1h_1} + 1\right)\right)^{\frac{1}{\left(h_1 - \alpha_{1h_1}\right)}}, \\
1 - y \geq \left(\Gamma\left(h_2 - \alpha_{2h_2} + 1\right)\right)^{\frac{1}{\left(h_2 - \alpha_{2h_2}\right)}}.
\end{array}
\right\}, \tag{6.43}
$$

equivalently,

$$\left\{\begin{array}{l} 0 < x \leq 1 - \left(\Gamma\left(h_1 - \alpha_{1h_1} + 1\right)\right)^{\frac{1}{\left(h_1 - \alpha_{1h_1}\right)}}, \\ 0 < y \leq 1 - \left(\Gamma\left(h_2 - \alpha_{2h_2} + 1\right)\right)^{\frac{1}{\left(h_2 - \alpha_{2h_2}\right)}}. \end{array}\right\}, \tag{6.44}$$

so again $\Psi \neq \varnothing$, non-trivially.

6.3 Main Results

We present our left side result:

Theorem 6.11 *Let h_1, h_2, v_1, v_2, r, p be integers, $0 \leq h_1 \leq v_1 \leq r$, $0 \leq h_2 \leq v_2 \leq p$ and let $f \in C^{r,p}\left([0,1]^2\right)$. Let $\alpha_{ij}(x, y)$, $i = h_1, h_1 + 1, ..., v_1$; $j = h_2, h_2 + 1, ..., v_2$ be real valued functions, defined and bounded in $[0, 1]^2$. We follow the terminology of section 6.2. We assume that $\alpha_{h_1h_2}$ is either $\geq \alpha > 0$ or $\leq \beta < 0$ throughout the critical region Φ, see (6.32). Let also $\alpha_{1i}, \alpha_{2j} \geq 0$, $\alpha_{1i}, \alpha_{2j} \notin \mathbb{Z}_+ : \lceil \alpha_{1i} \rceil = i$, $\lceil \alpha_{2j} \rceil = j$, $i = 0, 1, ..., r$; $j = 0, 1, ..., p$, $\alpha_{10} = 0$, $\alpha_{20} = 0$. Consider the bivariate left fractional differential operator*

$$L^* := \sum_{i=h_1}^{v_1} \sum_{j=h_2}^{v_2} \alpha_{ij}(x, y) D_{*0}^{(\alpha_{1i}, \alpha_{2j})}. \tag{6.45}$$

Let the integers $\overline{m_1}$, $\overline{m_2}$ with $\overline{m_1} > r$, $\overline{m_2} > p$. Set

$$l_{ij} := \left\| \alpha_{h_1h_2}^{-1}(x, y)\, \alpha_{ij}(x, y) \right\|_{\infty,[0,1]^2} < \infty. \tag{6.46}$$

Also set

$$M_{\overline{m_1},\overline{m_2}}^{i,j} := M_{\overline{m_1},\overline{m_2}}^{i,j}(f) := \lambda_{1i}\lambda_{2j}\left[t(i,j)\,\omega_1\left(f^{(i,j)}; \frac{1}{\sqrt{\overline{m_1} - i}}, \frac{1}{\sqrt{\overline{m_2} - j}}\right) + \max\left\{\frac{i(i-1)}{\overline{m_1}}, \frac{j(j-1)}{\overline{m_2}}\right\} \left\| f_{\infty,[0,1]^2}^{(i,j)} \right\|\right], \tag{6.47}$$

for $i = 0, 1, ..., r$; $j = 0, 1, ..., p$.

Above it is $t : \mathbb{Z}_+^2 \to \mathbb{R}_+$, and

$$\lambda_{1i} := \int_0^1 k_{1i}(z)\, dz, \quad \lambda_{2j} := \int_0^1 k_{2j}(z)\, dz \tag{6.48}$$

for $i = 1, ..., r$; $j = 1, ..., p$, and $\lambda_{10} := \lambda_{20} := 1$.

Call

$$P_{\overline{m_1},\overline{m_2}} := P_{\overline{m_1},\overline{m_2}}(f) := \sum_{i=h_1}^{v_1}\sum_{j=h_2}^{v_2} l_{ij} M^{i,j}_{\overline{m_1},\overline{m_2}}. \tag{6.49}$$

Then there exists a polynomial $Q_{\overline{m_1},\overline{m_2}}(x,y)$ of degree $(\overline{m_1},\overline{m_2})$ on $[0,1]^2$ such that
(i)

$$\left\| {}^{k_{1i}}_{k_{2j}}D^{(\alpha_{1i},\alpha_{2j})}_{*0}(f) - {}^{k_{1i}}_{k_{2j}}D^{(\alpha_{1i},\alpha_{2j})}_{*0}\left(Q_{\overline{m_1},\overline{m_2}}\right)\right\|_{\infty,[0,1]^2} \leq$$

$$\frac{P_{\overline{m_1},\overline{m_2}}}{h_1!h_2!}\left\| {}^{k_{1i}}_{k_{2j}}D^{(\alpha_{1i},\alpha_{2j})}_{*0}\left(x^{h_1}y^{h_2}\right)\right\|_{\infty,[0,1]^2} + M^{i,j}_{\overline{m_1},\overline{m_2}}, \tag{6.50}$$

for $(0,0) \leq (i,j) \leq (h_1,h_2)$,
and
(ii) if $(h_1+1, h_2+1) \leq (i,j) \leq (r,p)$, or $0 \leq i \leq h_1$, $h_2+1 \leq j \leq p$, or $h_1+1 \leq i \leq r$, $0 \leq j \leq h_2$,

$$\left\| {}^{k_{1i}}_{k_{2j}}D^{(\alpha_{1i},\alpha_{2j})}_{*0}(f) - {}^{k_{1i}}_{k_{2j}}D^{(\alpha_{1i},\alpha_{2j})}_{*0}\left(Q_{\overline{m_1},\overline{m_2}}\right)\right\|_{\infty,[0,1]^2} \leq M^{i,j}_{\overline{m_1},\overline{m_2}}. \tag{6.51}$$

By assuming $L^*f(x,y) \geq 0, \forall (x,y) \in \Phi$, we obtain $L^*\left(Q_{\overline{m_1},\overline{m_2}}(x,y)\right) \geq 0, \forall (x,y) \in \Phi$.

Proof We observe that ($\overline{m_1},\overline{m_2} \in \mathbb{N} : \overline{m_1} > i, \overline{m_2} > j$):

$$\left| {}^{k_{1i}}_{k_{2j}}D^{(\alpha_{1i},\alpha_{2j})}_{*0} f(x_1,x_2) - {}^{k_{1i}}_{k_{2j}}D^{(\alpha_{1i},\alpha_{2j})}_{*0}\left(B_{\overline{m_1},\overline{m_2}}f\right)(x_1,x_2)\right| =$$

$$\left| \int_0^{x_1}\int_0^{x_2} k_{1i}(x_1-t_1)\,k_{2j}(x_2-t_2)\frac{\partial^{i+j}f(t_1,t_2)}{\partial t_1^i\partial t_2^j}dt_1dt_2 - \right.$$

$$\left. \int_0^{x_1}\int_0^{x_2} k_{1i}(x_1-t_1)\,k_{2j}(x_2-t_2)\frac{\partial^{i+j}\left(B_{\overline{m_1},\overline{m_2}}f\right)(t_1,t_2)}{\partial t_1^i\partial t_2^j}dt_1dt_2\right| = \tag{6.52}$$

$$\left| \int_0^{x_1}\int_0^{x_2} k_{1i}(x_1-t_1)\,k_{2j}(x_2-t_2)\left(\frac{\partial^{i+j}f(t_1,t_2)}{\partial t_1^i\partial t_2^j} - \frac{\partial^{i+j}\left(B_{\overline{m_1},\overline{m_2}}f\right)(t_1,t_2)}{\partial t_1^i\partial t_2^j}\right)dt_1dt_2\right| \leq$$

$$\int_0^{x_1}\int_0^{x_2} k_{1i}(x_1-t_1)\,k_{2j}(x_2-t_2)\left|\frac{\partial^{i+j}f(t_1,t_2)}{\partial t_1^i\partial t_2^j} - \frac{\partial^{i+j}\left(B_{\overline{m_1},\overline{m_2}}f\right)(t_1,t_2)}{\partial t_1^i\partial t_2^j}\right|dt_1dt_2 \overset{(6.3)}{\leq}$$

$$\left(\int_0^{x_1}\int_0^{x_2} k_{1i}(x_1-t_1)\,k_{2j}(x_2-t_2)\,dt_1dt_2\right) \tag{6.53}$$

$$\left[t(i,j)\,\omega_1\left(f^{(i,j)};\frac{1}{\sqrt{\overline{m_1}-i}},\frac{1}{\sqrt{\overline{m_2}-j}}\right)+\max\left\{\frac{i(i-1)}{\overline{m_1}},\frac{j(j-1)}{\overline{m_2}}\right\}\left\|f^{(i,j)}\right\|_{\infty,[0,1]^2}\right]=$$

$$\left(\int_0^{x_1} k_{1i}(x_1-t_1)\,dt_1\right)\left(\int_0^{x_2} k_{2j}(x_2-t_2)\,dt_2\right)$$

$$\left[t(i,j)\,\omega_1\left(f^{(i,j)};\frac{1}{\sqrt{\overline{m_1}-i}},\frac{1}{\sqrt{\overline{m_2}-j}}\right)+\max\left\{\frac{i(i-1)}{\overline{m_1}},\frac{j(j-1)}{\overline{m_2}}\right\}\left\|f^{(i,j)}\right\|_{\infty,[0,1]^2}\right]=$$

$$\left(\int_0^{x_1} k_{1i}(z)\,dz\right)\left(\int_0^{x_2} k_{2j}(z)\,dz\right) \tag{6.54}$$

$$\left[t(i,j)\,\omega_1\left(f^{(i,j)};\frac{1}{\sqrt{\overline{m_1}-i}},\frac{1}{\sqrt{\overline{m_2}-j}}\right)+\max\left\{\frac{i(i-1)}{\overline{m_1}},\frac{j(j-1)}{\overline{m_2}}\right\}\left\|f^{(i,j)}\right\|_{\infty,[0,1]^2}\right]\leq$$

$$\left(\int_0^{1} k_{1i}(z)\,dz\right)\left(\int_0^{1} k_{2j}(z)\,dz\right)$$

$$\left[t(i,j)\,\omega_1\left(f^{(i,j)};\frac{1}{\sqrt{\overline{m_1}-i}},\frac{1}{\sqrt{\overline{m_2}-j}}\right)+\max\left\{\frac{i(i-1)}{\overline{m_1}},\frac{j(j-1)}{\overline{m_2}}\right\}\left\|f^{(i,j)}\right\|_{\infty,[0,1]^2}\right].$$

We have proved that

$$\left|{}^{k_{1i}}_{k_{2j}}D_{*0}^{(\alpha_{1i},\alpha_{2j})}f(x_1,x_2)-{}^{k_{1i}}_{k_{2j}}D_{*0}^{(\alpha_{1i},\alpha_{2j})}\left(B_{\overline{m_1},\overline{m_2}}f\right)(x_1,x_2)\right|\leq \tag{6.55}$$

$$\left(\int_0^{1} k_{1i}(z)\,dz\right)\left(\int_0^{1} k_{2j}(z)\,dz\right)$$

$$\left[t(i,j)\,\omega_1\left(f^{(i,j)};\frac{1}{\sqrt{\overline{m_1}-i}},\frac{1}{\sqrt{\overline{m_2}-j}}\right)+\max\left\{\frac{i(i-1)}{\overline{m_1}},\frac{j(j-1)}{\overline{m_2}}\right\}\left\|f^{(i,j)}\right\|_{\infty,[0,1]^2}\right],$$

$\forall\,(x_1,x_2)\in[0,1]^2$; $i=1,...,r$; $j=1,...,p$; where $\overline{m_1},\overline{m_2}\in\mathbb{N}:\overline{m_1}>r$, $\overline{m_2}>p$.

So we have established that

$$\left\| {}_{k_{2j}}^{k_{1i}} D_{*0}^{(\alpha_{1i},\alpha_{2j})} f(x_1,x_2) - {}_{k_{2j}}^{k_{1i}} D_{*0}^{(\alpha_{1i},\alpha_{2j})} \left(B_{\overline{m_1},\overline{m_2}} f\right)(x_1,x_2) \right\|_{\infty,[0,1]^2} \leq \tag{6.56}$$

$$\left(\int_0^1 k_{1i}(z)\,dz \right) \left(\int_0^1 k_{2j}(z)\,dz \right)$$

$$\left[t(i,j)\,\omega_1 \left(f^{(i,j)}; \frac{1}{\sqrt{\overline{m_1}-i}}, \frac{1}{\sqrt{\overline{m_2}-j}} \right) + \max\left\{ \frac{i(i-1)}{\overline{m_1}}, \frac{j(j-1)}{\overline{m_2}} \right\} \left\| f^{(i,j)} \right\|_{\infty,[0,1]^2} \right],$$

for $i = 1, ..., r$; $j = 1, ..., p$; $\overline{m_1}, \overline{m_2} \in \mathbb{N} : \overline{m_1} > r$, $\overline{m_2} > p$.

We call

$$\lambda_{1i} := \int_0^1 k_{1i}(z)\,dz, \;\; \lambda_{2j} := \int_0^1 k_{2j}(z)\,dz, \tag{6.57}$$

for $i = 1, ..., r$; $j = 1, ..., p$, as in (6.48).

We also set $\lambda_{10} := \lambda_{20} := 1$.

Thus, the following inequality is valid in general:

$$\left\| {}_{k_{2j}}^{k_{1i}} D_{*0}^{(\alpha_{1i},\alpha_{2j})} f(x_1,x_2) - {}_{k_{2j}}^{k_{1i}} D_{*0}^{(\alpha_{1i},\alpha_{2j})} \left(B_{\overline{m_1},\overline{m_2}} f\right)(x_1,x_2) \right\|_{\infty,[0,1]^2} \leq \lambda_{1i}\lambda_{2j}$$

$$\left[t(i,j)\,\omega_1 \left(f^{(i,j)}; \frac{1}{\sqrt{\overline{m_1}-i}}, \frac{1}{\sqrt{\overline{m_2}-j}} \right) + \max\left\{ \frac{i(i-1)}{\overline{m_1}}, \frac{j(j-1)}{\overline{m_2}} \right\} \left\| f^{(i,j)} \right\|_{\infty,[0,1]^2} \right]$$

$$= M^{i,j}_{\overline{m_1},\overline{m_2}}, \tag{6.58}$$

for $i = 0, 1, ..., r$; $j = 0, 1, ..., p$; $\overline{m_1}, \overline{m_2} \in \mathbb{N} : \overline{m_1} > r, \overline{m_2} > p$, case of Theorem 6.4.

Case (i) Assume throughout Φ that $\alpha_{h_1h_2}(x,y) \geq \alpha > 0$. Call

$$Q_{\overline{m_1},\overline{m_2}}(x,y) := B_{\overline{m_1},\overline{m_2}}(f;x,y) + P_{\overline{m_1},\overline{m_2}} \frac{x^{h_1}}{h_1!} \frac{y^{h_2}}{h_2!}, \tag{6.59}$$

$\forall (x,y) \in [0,1]^2$.

To remind, here it is

$$l_{ij} := \left\| \alpha_{h_1h_2}^{-1}(x,y)\,\alpha_{ij}(x,y) \right\|_{\infty,[0,1]^2}, \tag{6.60}$$

and

$$P_{\overline{m_1},\overline{m_2}} := \sum_{i=h_1}^{v_1} \sum_{j=h_2}^{v_2} l_{ij} M^{i,j}_{\overline{m_1},\overline{m_2}}.$$

Therefore by (6.58) we obtain

$$\left\| {}_{k_{2j}}^{k_{1i}}D_{*0}^{(\alpha_{1i},\alpha_{2j})}\left(f + P_{\overline{m_1},\overline{m_2}}\frac{x^{h_1}y^{h_2}}{h_1!h_2!}\right) - {}_{k_{2j}}^{k_{1i}}D_{*0}^{(\alpha_{1i},\alpha_{2j})}\left(Q_{\overline{m_1},\overline{m_2}}\right)\right\|_{\infty,[0,1]^2} \leq M_{\overline{m_1},\overline{m_2}}^{i,j}, \tag{6.61}$$

all $0 \leq i \leq r$, $0 \leq j \leq p$.

Let $(0,0) \leq (i,j) \leq (h_1,h_2)$, by (6.61) we get that

$$\left\| {}_{k_{2j}}^{k_{1i}}D_{*0}^{(\alpha_{1i},\alpha_{2j})}(f) - {}_{k_{2j}}^{k_{1i}}D_{*0}^{(\alpha_{1i},\alpha_{2j})}\left(Q_{\overline{m_1},\overline{m_2}}\right)\right\|_{\infty,[0,1]^2} \leq$$

$$\frac{P_{\overline{m_1},\overline{m_2}}}{h_1!h_2!}\left\| {}_{k_{2j}}^{k_{1i}}D_{*0}^{(\alpha_{1i},\alpha_{2j})}\left(x^{h_1}y^{h_2}\right)\right\|_{\infty,[0,1]^2} + M_{\overline{m_1},\overline{m_2}}^{i,j}, \tag{6.62}$$

proving (6.50).

If $(h_1+1,h_2+1) \leq (i,j) \leq (r,p)$, or $0 \leq i \leq h_1$, $h_2+1 \leq j \leq p$, or $h_1+1 \leq i \leq r$, $0 \leq j \leq h_2$ we get by (6.61) that

$$\left\| {}_{k_{2j}}^{k_{1i}}D_{*0}^{(\alpha_{1i},\alpha_{2j})}(f) - {}_{k_{2j}}^{k_{1i}}D_{*0}^{(\alpha_{1i},\alpha_{2j})}\left(Q_{\overline{m_1},\overline{m_2}}\right)\right\|_{\infty,[0,1]^2} \leq M_{\overline{m_1},\overline{m_2}}^{i,j}, \tag{6.63}$$

proving (6.51).

For (x,y) in the critical region Φ, we can write

$$\alpha_{h_1h_2}^{-1}(x,y)\,L^*\left(Q_{\overline{m_1},\overline{m_2}}(x,y)\right) = \alpha_{h_1h_2}^{-1}(x,y)\,L^*\left(f(x,y)\right) +$$

$$\frac{P_{\overline{m_1},\overline{m_2}}}{h_1!h_2!}\left({}_{k_{2h_2}}^{k_{1h_1}}D_{*0}^{(\alpha_{1h_1},\alpha_{2h_2})}\left(x^{h_1}y^{h_2}\right)\right) + \sum_{i=h_1}^{v_1}\sum_{j=h_2}^{v_2}\alpha_{h_1h_2}^{-1}(x,y)\,\alpha_{ij}(x,y)\;{}_{k_{2j}}^{k_{1i}}D_{*0}^{(\alpha_{1i},\alpha_{2j})} \tag{6.64}$$

$$\left[Q_{\overline{m_1},\overline{m_2}}(x,y) - f(x,y) - P_{\overline{m_1},\overline{m_2}}\frac{x^{h_1}}{h_1!}\frac{y^{h_2}}{h_2!}\right] \overset{(6.61)}{\geq}$$

(by $L^*f \geq 0$)

$$\frac{P_{\overline{m_1},\overline{m_2}}}{h_1!h_2!}\left({}_{k_{2h_2}}^{k_{1h_1}}D_{*0}^{(\alpha_{1h_1},\alpha_{2h_2})}\left(x^{h_1}y^{h_2}\right)\right) - \sum_{i=h_1}^{v_1}\sum_{j=h_2}^{v_2} l_{ij}M_{\overline{m_1},\overline{m_2}}^{i,j} =$$

$$P_{\overline{m_1},\overline{m_2}}\left[\frac{{}_{k_{2h_2}}^{k_{1h_1}}D_{*0}^{(\alpha_{1h_1},\alpha_{2h_2})}\left(x^{h_1}y^{h_2}\right)}{h_1!h_2!} - 1\right] =$$

$$P_{\overline{m_1},\overline{m_2}}\left[\frac{\left({}_{k_{20}}^{k_{1h_1}}D_{*0}^{(\alpha_{1h_1},0)}x^{h_1}\right)\left({}_{k_{2h_2}}^{k_{10}}D_{*0}^{(0,\alpha_{2h_2})}y^{h_2}\right)}{h_1!h_2!}-1\right]=:\varphi. \quad (6.65)$$

If $h_1 = h_2 = 0$, then $\alpha_{1h_1} = \alpha_{2h_2} = 0$, and $\varphi = 0$.
If $h_1 = 0$, and $h_2 \neq 0$, then

$$\varphi = P_{\overline{m_1},\overline{m_2}}\left[\frac{{}_{k_{2h_2}}^{k_{10}}D_{*0}^{(0,\alpha_{2h_2})}y^{h_2}}{h_2!}-1\right]=$$

$$P_{\overline{m_1},\overline{m_2}}\left[\int_0^y k_{2h_2}(y-t)\,dt-1\right]=P_{\overline{m_1},\overline{m_2}}\left(\int_0^y k_{2h_2}(z)\,dz-1\right)\geq 0. \quad (6.66)$$

Similarly, we treat the case $h_1 \neq 0$, $h_2 = 0$.
When $h_1, h_2 \neq 0$, then we have

$$\varphi = P_{\overline{m_1},\overline{m_2}}\left[\left(\int_0^x k_{1h_1}(x-t)\,dt\right)\left(\int_0^y k_{2h_2}(y-t)\,dt\right)-1\right]=$$

$$P_{\overline{m_1},\overline{m_2}}\left[\left(\int_0^x k_{1h_1}(z)\,dz\right)\left(\int_0^y k_{2h_2}(z)\,dz\right)-1\right]\geq 0. \quad (6.67)$$

So in all four cases we have proved that

$$L^*\left(Q_{\overline{m_1},\overline{m_2}}\right)(x,y)\geq 0,\ \forall\ (x,y)\in\Phi. \quad (6.68)$$

Case (ii) Assume throughout Φ that $\alpha_{h_1h_2} \leq \beta < 0$. Consider

$$\overline{Q}_{\overline{m_1},\overline{m_2}}(x,y):=B_{\overline{m_1},\overline{m_2}}(f;x,y)-P_{\overline{m_1},\overline{m_2}}\frac{x^{h_1}}{h_1!}\frac{y^{h_2}}{h_2!},$$

$\forall\ (x,y)\in[0,1]^2$.
Therefore by (6.58) we obtain

$$\left\|{}_{k_{2j}}^{k_{1i}}D_{*0}^{(\alpha_{1i},\alpha_{2j})}\left(f-P_{\overline{m_1},\overline{m_2}}\frac{x^{h_1}y^{h_2}}{h_1!h_2!}\right)-{}_{k_{2j}}^{k_{1i}}D_{*0}^{(\alpha_{1i},\alpha_{2j})}\left(\overline{Q}_{\overline{m_1},\overline{m_2}}\right)\right\|_{\infty,[0,1]^2}\leq M_{\overline{m_1},\overline{m_2}}^{i,j}, \quad (6.69)$$

all $0 \leq i \leq r$, $0 \leq j \leq p$.
Also both (6.50), (6.51) are valid, just replace $Q_{\overline{m_1},\overline{m_2}}$ by $\overline{Q}_{\overline{m_1},\overline{m_2}}$ in (6.62), (6.63).
For (x, y) in the critical region Φ, we can write

$$\alpha_{h_1h_2}^{-1}(x,y)\,L^*\left(\overline{Q}_{\overline{m_1},\overline{m_2}}(x,y)\right)=\alpha_{h_1h_2}^{-1}(x,y)\,L^*(f(x,y))-$$

$$\frac{P_{\overline{m_1},\overline{m_2}}}{h_1!h_2!}\left({}_{k_{2h_2}}^{k_{1h_1}}D_{*0}^{(\alpha_{1h_1},\alpha_{2h_2})}\left(x^{h_1}y^{h_2}\right)\right)+\sum_{i=h_1}^{\nu_1}\sum_{j=h_2}^{\nu_2}\alpha_{h_1h_2}^{-1}(x,y)\,\alpha_{ij}(x,y)\,{}_{k_{2j}}^{k_{1i}}D_{*0}^{(\alpha_{1i},\alpha_{2j})}$$

$$\left[\overline{Q}_{\overline{m_1},\overline{m_2}}(x,y)-f(x,y)+P_{\overline{m_1},\overline{m_2}}\frac{x^{h_1}}{h_1!}\frac{y^{h_2}}{h_2!}\right]\overset{(6.69)}{\leq}\tag{6.70}$$

(by $L^*f\geq 0$)

$$-\frac{P_{\overline{m_1},\overline{m_2}}}{h_1!h_2!}\left({}_{k_{2h_2}}^{k_{1h_1}}D_{*0}^{(\alpha_{1h_1},\alpha_{2h_2})}\left(x^{h_1}y^{h_2}\right)\right)+\sum_{i=h_1}^{\nu_1}\sum_{j=h_2}^{\nu_2}l_{ij}M_{\overline{m_1},\overline{m_2}}^{i,j}=$$

$$P_{\overline{m_1},\overline{m_2}}\left[1-\frac{{}_{k_{2h_2}}^{k_{1h_1}}D_{*0}^{(\alpha_{1h_1},\alpha_{2h_2})}\left(x^{h_1}y^{h_2}\right)}{h_1!h_2!}\right]=$$

$$P_{\overline{m_1},\overline{m_2}}\left[1-\frac{\left({}_{k_{20}}^{k_{1h_1}}D_{*0}^{(\alpha_{1h_1},0)}x^{h_1}\right)\left({}_{k_{2h_2}}^{k_{10}}D_{*0}^{(0,\alpha_{2h_2})}y^{h_2}\right)}{h_1!h_2!}\right]=:\psi.\tag{6.71}$$

If $h_1=h_2=0$, then $\alpha_{1h_1}=\alpha_{2h_2}=0$, and $\psi=0$.
If $h_1=0$, and $h_2\neq 0$, then

$$\varphi=P_{\overline{m_1},\overline{m_2}}\left[1-\frac{{}_{k_{2h_2}}^{k_{10}}D_{*0}^{(0,\alpha_{2h_2})}y^{h_2}}{h_2!}\right]=$$

$$P_{\overline{m_1},\overline{m_2}}\left[1-\int_0^y k_{2h_2}(y-t)\,dt\right]=P_{\overline{m_1},\overline{m_2}}\left[1-\int_0^y k_{2h_2}(z)\,dz\right]\leq 0.\tag{6.72}$$

Similarly, we treat the case $h_1\neq 0$, $h_2=0$.

When $h_1,h_2\neq 0$, then we have

$$\varphi=P_{\overline{m_1},\overline{m_2}}\left[1-\left(\int_0^x k_{1h_1}(x-t)\,dt\right)\left(\int_0^y k_{2h_2}(y-t)\,dt\right)\right]=$$

$$P_{\overline{m_1},\overline{m_2}}\left[1-\left(\int_0^x k_{1h_1}(z)\,dz\right)\left(\int_0^y k_{2h_2}(z)\,dz\right)\right]\leq 0.\tag{6.73}$$

So in all four cases we have proved that

$$L^* \left(\overline{Q}_{\overline{m_1},\overline{m_2}}\right)(x, y) \geq 0, \ \forall \ (x, y) \in \Phi. \tag{6.74}$$

■

Next we give our right side result:

Theorem 6.12 *Let h_1, h_2 be even and v_1, v_2, r, p be integers, $0 \leq h_1 \leq v_1 \leq r$, $0 \leq h_2 \leq v_2 \leq p$ and let $f \in C^{r,p}\left([0,1]^2\right)$. Let $\alpha_{ij}(x, y)$, $i = h_1, h_1 + 1, ..., v_1$; $j = h_2, h_2 + 1, ..., v_2$ be real valued functions, defined and bounded in $[0, 1]^2$. We follow the terminology of section 6.2. We assume that $\alpha_{h_1h_2}$ is either $\geq \alpha > 0$ or $\leq \beta < 0$ throughout the critical region Ψ, see (6.42). Let also $\alpha_{1i}, \alpha_{2j} \geq 0$, $\alpha_{1i}, \alpha_{2j} \notin \mathbb{Z}_+$: $\lceil \alpha_{1i} \rceil = i$, $\lceil \alpha_{2j} \rceil = j$, $i = 0, 1, ..., r$; $j = 0, 1, ..., p$, $\alpha_{10} = 0$, $\alpha_{20} = 0$. Consider the bivariate right fractional differential operator*

$$L^* := \sum_{i=h_1}^{v_1} \sum_{j=h_2}^{v_2} \alpha_{ij}(x, y) D_{1-}^{(\alpha_{1i},\alpha_{2j})}. \tag{6.75}$$

Let the integers $\overline{m_1}, \overline{m_2}$ with $\overline{m_1} > r$, $\overline{m_2} > p$. Set

$$l_{ij} := \left\| \alpha_{h_1h_2}^{-1}(x, y)\, \alpha_{ij}(x, y) \right\|_{\infty,[0,1]^2} < \infty. \tag{6.76}$$

Also set

$$M_{\overline{m_1},\overline{m_2}}^{i,j} := M_{\overline{m_1},\overline{m_2}}^{i,j}(f) := \lambda_{1i}\lambda_{2j}\left[t(i, j)\, \omega_1\left(f^{(i,j)}; \frac{1}{\sqrt{\overline{m_1} - i}}, \frac{1}{\sqrt{\overline{m_2} - j}}\right) + \right.$$

$$\left. \max\left\{\frac{i(i-1)}{\overline{m_1}}, \frac{j(j-1)}{\overline{m_2}}\right\} \left\|f_{\infty,[0,1]^2}^{(i,j)}\right\|\right], \tag{6.77}$$

for $i = 0, 1, ..., r$; $j = 0, 1, ..., p$.
Above it is $t : \mathbb{Z}_+^2 \to \mathbb{R}_+$, and

$$\lambda_{1i} := \int_0^1 k_{1i}(z)\, dz, \quad \lambda_{2j} := \int_0^1 k_{2j}(z)\, dz \tag{6.78}$$

for $i = 1, ..., r$; $j = 1, ..., p$, and $\lambda_{10} := \lambda_{20} := 1$.
Call

$$P_{\overline{m_1},\overline{m_2}} := P_{\overline{m_1},\overline{m_2}}(f) := \sum_{i=h_1}^{v_1} \sum_{j=h_2}^{v_2} l_{ij} M_{\overline{m_1},\overline{m_2}}^{i,j}. \tag{6.79}$$

Then there exists a polynomial $Q_{\overline{m_1},\overline{m_2}}(x, y)$ of degree $(\overline{m_1}, \overline{m_2})$ on $[0, 1]^2$ such that

(i)

$$\left\| {}_{k_{2j}}^{k_{1i}}D_{1-}^{(\alpha_{1i},\alpha_{2j})}(f) - {}_{k_{2j}}^{k_{1i}}D_{1-}^{(\alpha_{1i},\alpha_{2j})}\left(Q_{\overline{m_1},\overline{m_2}}\right)\right\|_{\infty,[0,1]^2} \leq$$

$$\frac{P_{\overline{m_1},\overline{m_2}}}{h_1!h_2!}\left\| {}_{k_{2j}}^{k_{1i}}D_{1-}^{(\alpha_{1i},\alpha_{2j})}\left(x^{h_1}y^{h_2}\right)\right\|_{\infty,[0,1]^2} + M_{\overline{m_1},\overline{m_2}}^{i,j}, \tag{6.80}$$

for $(0,0) \leq (i,j) \leq (h_1,h_2)$,
and

(ii) if $(h_1+1,h_2+1) \leq (i,j) \leq (r,p)$, *or* $0 \leq i \leq h_1$, $h_2+1 \leq j \leq p$, *or* $h_1 + 1 \leq i \leq r$, $0 \leq j \leq h_2$,

$$\left\| {}_{k_{2j}}^{k_{1i}}D_{1-}^{(\alpha_{1i},\alpha_{2j})}(f) - {}_{k_{2j}}^{k_{1i}}D_{1-}^{(\alpha_{1i},\alpha_{2j})}\left(Q_{\overline{m_1},\overline{m_2}}\right)\right\|_{\infty,[0,1]^2} \leq M_{\overline{m_1},\overline{m_2}}^{i,j}. \tag{6.81}$$

By assuming $L^*f(x,y) \geq 0, \forall (x,y) \in \Psi$, *we obtain* $L^*\left(Q_{\overline{m_1},\overline{m_2}}(x,y)\right) \geq 0, \forall (x,y) \in \Psi$.

Proof We observe that ($\overline{m_1},\overline{m_2} \in \mathbb{N} : \overline{m_1} > i, \overline{m_2} > j$):

$$\left| {}_{k_{2j}}^{k_{1i}}D_{1-}^{(\alpha_{1i},\alpha_{2j})}f(x_1,x_2) - {}_{k_{2j}}^{k_{1i}}D_{1-}^{(\alpha_{1i},\alpha_{2j})}\left(B_{\overline{m_1},\overline{m_2}}f\right)(x_1,x_2)\right| =$$

$$\left|(-1)^{i+j}\int_{x_1}^{1}\int_{x_2}^{1} k_{1i}(t_1-x_1)\,k_{2j}(t_2-x_2)\,\frac{\partial^{i+j}f(t_1,t_2)}{\partial t_1^i\partial t_2^j}dt_1dt_2 -\right.$$

$$\left.(-1)^{i+j}\int_{x_1}^{1}\int_{x_2}^{1} k_{1i}(t_1-x_1)\,k_{2j}(t_2-x_2)\,\frac{\partial^{i+j}\left(B_{\overline{m_1},\overline{m_2}}f\right)(t_1,t_2)}{\partial t_1^i\partial t_2^j}dt_1dt_2\right| = \tag{6.82}$$

$$\left|\int_{x_1}^{1}\int_{x_2}^{1} k_{1i}(t_1-x_1)\,k_{2j}(t_2-x_2)\left(\frac{\partial^{i+j}f(t_1,t_2)}{\partial t_1^i\partial t_2^j} - \frac{\partial^{i+j}\left(B_{\overline{m_1},\overline{m_2}}f\right)(t_1,t_2)}{\partial t_1^i\partial t_2^j}\right)dt_1dt_2\right| \leq$$

$$\int_{x_1}^{1}\int_{x_2}^{1} k_{1i}(t_1-x_1)\,k_{2j}(t_2-x_2)\left|\frac{\partial^{i+j}f(t_1,t_2)}{\partial t_1^i\partial t_2^j} - \frac{\partial^{i+j}\left(B_{\overline{m_1},\overline{m_2}}f\right)(t_1,t_2)}{\partial t_1^i\partial t_2^j}\right|dt_1dt_2 \overset{(6.3)}{\leq}$$

$$\left(\int_{x_1}^{1}\int_{x_2}^{1} k_{1i}(t_1-x_1)\,k_{2j}(t_2-x_2)\,dt_1dt_2\right) \tag{6.83}$$

$$\left[t(i,j)\,\omega_1\left(f^{(i,j)};\frac{1}{\sqrt{\overline{m_1}-i}},\frac{1}{\sqrt{\overline{m_2}-j}}\right) + \max\left\{\frac{i(i-1)}{\overline{m_1}},\frac{j(j-1)}{\overline{m_2}}\right\}\left\|f^{(i,j)}\right\|_{\infty,[0,1]^2}\right] =$$

$$\left(\int_{x_1}^{1} k_{1i}(t_1 - x_1)\, dt_1\right)\left(\int_{x_2}^{1} k_{2j}(t_2 - x_2)\, dt_2\right)$$

$$\left[t(i,j)\,\omega_1\left(f^{(i,j)}; \frac{1}{\sqrt{\overline{m_1} - i}}, \frac{1}{\sqrt{\overline{m_2} - j}}\right) + \max\left\{\frac{i(i-1)}{\overline{m_1}}, \frac{j(j-1)}{\overline{m_2}}\right\}\left\|f^{(i,j)}\right\|_{\infty,[0,1]^2}\right] =$$

$$\left(\int_{0}^{1-x_1} k_{1i}(z)\, dz\right)\left(\int_{0}^{1-x_2} k_{2j}(z)\, dz\right) \tag{6.84}$$

$$\left[t(i,j)\,\omega_1\left(f^{(i,j)}; \frac{1}{\sqrt{\overline{m_1} - i}}, \frac{1}{\sqrt{\overline{m_2} - j}}\right) + \max\left\{\frac{i(i-1)}{\overline{m_1}}, \frac{j(j-1)}{\overline{m_2}}\right\}\left\|f^{(i,j)}\right\|_{\infty,[0,1]^2}\right] \leq$$

$$\left(\int_{0}^{1} k_{1i}(z)\, dz\right)\left(\int_{0}^{1} k_{2j}(z)\, dz\right)$$

$$\left[t(i,j)\,\omega_1\left(f^{(i,j)}; \frac{1}{\sqrt{\overline{m_1} - i}}, \frac{1}{\sqrt{\overline{m_2} - j}}\right) + \max\left\{\frac{i(i-1)}{\overline{m_1}}, \frac{j(j-1)}{\overline{m_2}}\right\}\left\|f^{(i,j)}\right\|_{\infty,[0,1]^2}\right].$$

We have proved that

$$\left|{}_{k_{2j}}^{k_{1i}}D_{1-}^{(\alpha_{1i},\alpha_{2j})} f(x_1, x_2) - {}_{k_{2j}}^{k_{1i}}D_{1-}^{(\alpha_{1i},\alpha_{2j})}\left(B_{\overline{m_1},\overline{m_2}} f\right)(x_1, x_2)\right| \leq \tag{6.85}$$

$$\left(\int_{0}^{1} k_{1i}(z)\, dz\right)\left(\int_{0}^{1} k_{2j}(z)\, dz\right)$$

$$\left[t(i,j)\,\omega_1\left(f^{(i,j)}; \frac{1}{\sqrt{\overline{m_1} - i}}, \frac{1}{\sqrt{\overline{m_2} - j}}\right) + \max\left\{\frac{i(i-1)}{\overline{m_1}}, \frac{j(j-1)}{\overline{m_2}}\right\}\left\|f^{(i,j)}\right\|_{\infty,[0,1]^2}\right],$$

$\forall\, (x_1, x_2) \in [0,1]^2$; $i = 1, ..., r$; $j = 1, ..., p$; where $\overline{m_1}, \overline{m_2} \in \mathbb{N} : \overline{m_1} > r$, $\overline{m_2} > p$.

So we have established that

$$\left\|{}_{k_{2j}}^{k_{1i}}D_{1-}^{(\alpha_{1i},\alpha_{2j})} f(x_1, x_2) - {}_{k_{2j}}^{k_{1i}}D_{1-}^{(\alpha_{1i},\alpha_{2j})}\left(B_{\overline{m_1},\overline{m_2}} f\right)(x_1, x_2)\right\|_{\infty,[0,1]^2} \leq \tag{6.86}$$

$$\left(\int_{0}^{1} k_{1i}(z)\, dz\right)\left(\int_{0}^{1} k_{2j}(z)\, dz\right)$$

$$\left[t(i,j)\,\omega_1\left(f^{(i,j)};\frac{1}{\sqrt{\overline{m_1}-i}},\frac{1}{\sqrt{\overline{m_2}-j}}\right)+\max\left\{\frac{i(i-1)}{\overline{m_1}},\frac{j(j-1)}{\overline{m_2}}\right\}\left\|f^{(i,j)}\right\|_{\infty,[0,1]^2}\right],$$

for $i=1,...,r$; $j=1,...,p$; $\overline{m_1},\overline{m_2}\in\mathbb{N}:\overline{m_1}>r,\overline{m_2}>p$.

We call

$$\lambda_{1i}:=\int_0^1 k_{1i}(z)\,dz,\ \lambda_{2j}:=\int_0^1 k_{2j}(z)\,dz, \tag{6.87}$$

for $i=1,...,r$; $j=1,...,p$, as in (6.78).

We also set $\lambda_{10}:=\lambda_{20}:=1$.

Thus, the following inequality is valid in general:

$$\left\|{}_{k_{2j}}^{k_{1i}}D_{1-}^{(\alpha_{1i},\alpha_{2j})}f(x_1,x_2)-{}_{k_{2j}}^{k_{1i}}D_{1-}^{(\alpha_{1i},\alpha_{2j})}\left(B_{\overline{m_1},\overline{m_2}}f\right)(x_1,x_2)\right\|_{\infty,[0,1]^2}\leq\lambda_{1i}\lambda_{2j}$$

$$\left[t(i,j)\,\omega_1\left(f^{(i,j)};\frac{1}{\sqrt{\overline{m_1}-i}},\frac{1}{\sqrt{\overline{m_2}-j}}\right)+\max\left\{\frac{i(i-1)}{\overline{m_1}},\frac{j(j-1)}{\overline{m_2}}\right\}\left\|f^{(i,j)}\right\|_{\infty,[0,1]^2}\right]$$

$$=M_{\overline{m_1},\overline{m_2}}^{i,j}, \tag{6.88}$$

for $i=0,1,...,r$; $j=0,1,...,p$; $\overline{m_1},\overline{m_2}\in\mathbb{N}:\overline{m_1}>r,\overline{m_2}>p$, case of Theorem 6.4.

Case (i) Assume throughout Ψ that $\alpha_{h_1h_2}(x,y)\geq\alpha>0$. Call

$$Q_{\overline{m_1},\overline{m_2}}(x,y):=B_{\overline{m_1},\overline{m_2}}(f;x,y)+P_{\overline{m_1},\overline{m_2}}\frac{x^{h_1}}{h_1!}\frac{y^{h_2}}{h_2!}, \tag{6.89}$$

$\forall\,(x,y)\in[0,1]^2$.

To remind, here it is

$$l_{ij}:=\left\|\alpha_{h_1h_2}^{-1}(x,y)\,\alpha_{ij}(x,y)\right\|_{\infty,[0,1]^2}, \tag{6.90}$$

and

$$P_{\overline{m_1},\overline{m_2}}:=\sum_{i=h_1}^{v_1}\sum_{j=h_2}^{v_2}l_{ij}M_{\overline{m_1},\overline{m_2}}^{i,j}.$$

Therefore by (6.88) we obtain

$$\left\|{}_{k_{2j}}^{k_{1i}}D_{1-}^{(\alpha_{1i},\alpha_{2j})}\left(f+P_{\overline{m_1},\overline{m_2}}\frac{x^{h_1}y^{h_2}}{h_1!h_2!}\right)-{}_{k_{2j}}^{k_{1i}}D_{1-}^{(\alpha_{1i},\alpha_{2j})}\left(Q_{\overline{m_1},\overline{m_2}}\right)\right\|_{\infty,[0,1]^2}\leq M_{\overline{m_1},\overline{m_2}}^{i,j}, \tag{6.91}$$

all $0\leq i\leq r$, $0\leq j\leq p$.

Let $(0,0) \leq (i,j) \leq (h_1,h_2)$, by (6.91) we get that

$$\left\| {}^{k_{1i}}_{k_{2j}}D^{(\alpha_{1i},\alpha_{2j})}_{1-}(f) - {}^{k_{1i}}_{k_{2j}}D^{(\alpha_{1i},\alpha_{2j})}_{1-}\left(Q_{\overline{m_1},\overline{m_2}}\right)\right\|_{\infty,[0,1]^2} \leq$$

$$\frac{P_{\overline{m_1},\overline{m_2}}}{h_1!h_2!}\left\| {}^{k_{1i}}_{k_{2j}}D^{(\alpha_{1i},\alpha_{2j})}_{1-}\left(x^{h_1}y^{h_2}\right)\right\|_{\infty,[0,1]^2} + M^{i,j}_{\overline{m_1},\overline{m_2}}, \tag{6.92}$$

proving (6.80).

If $(h_1+1,h_2+1) \leq (i,j) \leq (r,p)$, or $0 \leq i \leq h_1$, $h_2+1 \leq j \leq p$, or $h_1+1 \leq i \leq r$, $0 \leq j \leq h_2$ we get by (6.91) that

$$\left\| {}^{k_{1i}}_{k_{2j}}D^{(\alpha_{1i},\alpha_{2j})}_{1-}(f) - {}^{k_{1i}}_{k_{2j}}D^{(\alpha_{1i},\alpha_{2j})}_{1-}\left(Q_{\overline{m_1},\overline{m_2}}\right)\right\|_{\infty,[0,1]^2} \leq M^{i,j}_{\overline{m_1},\overline{m_2}}, \tag{6.93}$$

proving (6.81).

For (x,y) in the critical region Ψ, we can write

$$\alpha^{-1}_{h_1h_2}(x,y)\,L^*\left(Q_{\overline{m_1},\overline{m_2}}(x,y)\right) = \alpha^{-1}_{h_1h_2}(x,y)\,L^*\left(f(x,y)\right) +$$

$$\frac{P_{\overline{m_1},\overline{m_2}}}{h_1!h_2!}\left({}^{k_{1h_1}}_{k_{2h_2}}D^{(\alpha_{1h_1},\alpha_{2h_2})}_{1-}\left(x^{h_1}y^{h_2}\right)\right) + \sum_{i=h_1}^{v_1}\sum_{j=h_2}^{v_2}\alpha^{-1}_{h_1h_2}(x,y)\,\alpha_{ij}(x,y)\,{}^{k_{1i}}_{k_{2j}}D^{(\alpha_{1i},\alpha_{2j})}_{1-} \tag{6.94}$$

$$\left[Q_{\overline{m_1},\overline{m_2}}(x,y) - f(x,y) - P_{\overline{m_1},\overline{m_2}}\frac{x^{h_1}}{h_1!}\frac{y^{h_2}}{h_2!}\right] \overset{(6.91)}{\geq}$$

(by $L^*f \geq 0$)

$$\frac{P_{\overline{m_1},\overline{m_2}}}{h_1!h_2!}\left({}^{k_{1h_1}}_{k_{2h_2}}D^{(\alpha_{1h_1},\alpha_{2h_2})}_{1-}\left(x^{h_1}y^{h_2}\right)\right) - \sum_{i=h_1}^{v_1}\sum_{j=h_2}^{v_2} l_{ij}M^{i,j}_{\overline{m_1},\overline{m_2}} =$$

$$P_{\overline{m_1},\overline{m_2}}\left[\frac{{}^{k_{1h_1}}_{k_{2h_2}}D^{(\alpha_{1h_1},\alpha_{2h_2})}_{1-}\left(x^{h_1}y^{h_2}\right)}{h_1!h_2!} - 1\right] =$$

$$P_{\overline{m_1},\overline{m_2}}\left[\frac{\left({}^{k_{1h_1}}_{k_{20}}D^{(\alpha_{1h_1},0)}_{1-}x^{h_1}\right)\left({}^{k_{10}}_{k_{2h_2}}D^{(0,\alpha_{2h_2})}_{1-}y^{h_2}\right)}{h_1!h_2!} - 1\right] =: \psi. \tag{6.95}$$

If $h_1 = h_2 = 0$, then $\alpha_{1h_1} = \alpha_{2h_2} = 0$, and $\psi = 0$.

If $h_1 = 0$, and $h_2 \neq 0$, then

$$\psi = P_{\overline{m_1},\overline{m_2}} \left[\frac{{}_{k_{2h_2}}^{k_{10}} D_{1-}^{(0,\alpha_{2h_2})} y^{h_2}}{h_2!} - 1 \right] =$$

$$P_{\overline{m_1},\overline{m_2}} \left[(-1)^{h_2} \int_y^1 k_{2h_2}(t-y)\,dt - 1 \right] \overset{(h_2 \text{ is even})}{=}$$

$$P_{\overline{m_1},\overline{m_2}} \left(\int_0^{1-y} k_{2h_2}(z)\,dz - 1 \right) \geq 0. \quad (6.96)$$

Similarly, we treat the case $h_1 \neq 0$, $h_2 = 0$.

When $h_1, h_2 \neq 0$, then we have

$$\psi = P_{\overline{m_1},\overline{m_2}} \left[\left(\int_x^1 k_{1h_1}(t-x)\,dt \right) \left(\int_y^1 k_{2h_2}(t-y)\,dt \right) - 1 \right] =$$

$$P_{\overline{m_1},\overline{m_2}} \left[\left(\int_0^{1-x} k_{1h_1}(z)\,dz \right) \left(\int_0^{1-y} k_{2h_2}(z)\,dz \right) - 1 \right] \geq 0. \quad (6.97)$$

So in all four cases we have proved that

$$L^* \left(Q_{\overline{m_1},\overline{m_2}} \right)(x,y) \geq 0, \ \forall\ (x,y) \in \Psi. \quad (6.98)$$

Case (ii) Assume throughout Ψ that $\alpha_{h_1 h_2} \leq \beta < 0$. Consider

$$\overline{Q}_{\overline{m_1},\overline{m_2}}(x,y) := B_{\overline{m_1},\overline{m_2}}(f; x,y) - P_{\overline{m_1},\overline{m_2}} \frac{x^{h_1}}{h_1!} \frac{y^{h_2}}{h_2!},$$

$\forall\ (x,y) \in [0,1]^2$.

Therefore by (6.88) we obtain

$$\left\| {}_{k_{2j}}^{k_{1i}} D_{1-}^{(\alpha_{1i},\alpha_{2j})} \left(f - P_{\overline{m_1},\overline{m_2}} \frac{x^{h_1} y^{h_2}}{h_1! h_2!} \right) - {}_{k_{2j}}^{k_{1i}} D_{1-}^{(\alpha_{1i},\alpha_{2j})} \left(\overline{Q}_{\overline{m_1},\overline{m_2}} \right) \right\|_{\infty,[0,1]^2} \leq M_{\overline{m_1},\overline{m_2}}^{i,j}, \quad (6.99)$$

all $0 \leq i \leq r$, $0 \leq j \leq p$.

Also both (6.80), (6.81) are valid, just replace $Q_{\overline{m_1},\overline{m_2}}$ by $\overline{Q}_{\overline{m_1},\overline{m_2}}$ in (6.92), (6.93).

For (x,y) in the critical region Ψ, we can write

$$\alpha_{h_1h_2}^{-1}(x,y)\,L^*\left(\overline{Q}_{\overline{m_1},\overline{m_2}}(x,y)\right)=\alpha_{h_1h_2}^{-1}(x,y)\,L^*\left(f(x,y)\right)-$$

$$\frac{P_{\overline{m_1},\overline{m_2}}}{h_1!h_2!}\left({}_{k_{2h_2}}^{k_{1h_1}}D_{1-}^{(\alpha_{1h_1},\alpha_{2h_2})}\left(x^{h_1}y^{h_2}\right)\right)+\sum_{i=h_1}^{\nu_1}\sum_{j=h_2}^{\nu_2}\alpha_{h_1h_2}^{-1}(x,y)\,\alpha_{ij}(x,y)\;{}_{k_{2j}}^{k_{1i}}D_{1-}^{(\alpha_{1i},\alpha_{2j})}$$

$$\left[\overline{Q}_{\overline{m_1},\overline{m_2}}(x,y)-f(x,y)+P_{\overline{m_1},\overline{m_2}}\frac{x^{h_1}}{h_1!}\frac{y^{h_2}}{h_2!}\right]\overset{(6.99)}{\leq}\tag{6.100}$$

(by $L^*f\geq 0$)

$$-\frac{P_{\overline{m_1},\overline{m_2}}}{h_1!h_2!}\left({}_{k_{2h_2}}^{k_{1h_1}}D_{1-}^{(\alpha_{1h_1},\alpha_{2h_2})}\left(x^{h_1}y^{h_2}\right)\right)+\sum_{i=h_1}^{\nu_1}\sum_{j=h_2}^{\nu_2}l_{ij}M_{\overline{m_1},\overline{m_2}}^{i,j}=$$

$$P_{\overline{m_1},\overline{m_2}}\left[1-\frac{{}_{k_{2h_2}}^{k_{1h_1}}D_{1-}^{(\alpha_{1h_1},\alpha_{2h_2})}\left(x^{h_1}y^{h_2}\right)}{h_1!h_2!}\right]=$$

$$P_{\overline{m_1},\overline{m_2}}\left[1-\frac{\left({}_{k_{20}}^{k_{1h_1}}D_{1-}^{(\alpha_{1h_1},0)}x^{h_1}\right)\left({}_{k_{2h_2}}^{k_{10}}D_{1-}^{(0,\alpha_{2h_2})}y^{h_2}\right)}{h_1!h_2!}\right]=:\psi.\tag{6.101}$$

If $h_1=h_2=0$, then $\alpha_{1h_1}=\alpha_{2h_2}=0$, and $\psi=0$.
If $h_1=0$, and $h_2\neq 0$, then

$$\psi=P_{\overline{m_1},\overline{m_2}}\left[1-\frac{{}_{k_{2h_2}}^{k_{10}}D_{1-}^{(0,\alpha_{2h_2})}y^{h_2}}{h_2!}\right]=$$

$$P_{\overline{m_1},\overline{m_2}}\left[1-\int_y^1 k_{2h_2}(t-y)\,dt\right]=P_{\overline{m_1},\overline{m_2}}\left[1-\int_0^{1-y}k_{2h_2}(z)\,dz\right]\leq 0.\tag{6.102}$$

Similarly, we treat the case $h_1\neq 0$, $h_2=0$.
When $h_1,h_2\neq 0$, then we have

$$\psi=P_{\overline{m_1},\overline{m_2}}\left[1-\left(\int_x^1 k_{1h_1}(t-x)\,dt\right)\left(\int_y^1 k_{2h_2}(t-y)\,dt\right)\right]=\tag{6.103}$$

$$P_{\overline{m_1},\overline{m_2}}\left[1-\left(\int_0^{1-x}k_{1h_1}(z)\,dz\right)\left(\int_0^{1-y}k_{2h_2}(z)\,dz\right)\right]\leq 0.$$

So in all four cases we have proved that

$$L^* \left(\overline{Q_{\overline{m_1},\overline{m_2}}}\right)(x, y) \geq 0, \ \forall \ (x, y) \in \Psi. \tag{6.104}$$

■

Conclusion 6.13 Theorem 6.11 generalizes greatly Theorems 6.9, and 6.12 generalizes greatly the main Theorem 2.9 of [3]. These generalizations involve abstract fractional kernels that can be singular or non-singular kernels. That is we cover all kinds of left and right Caputo type fractional calculi of singular and non-singular kernel.

References

1. Anastassiou, G.A.: Bivariate monotone approximation. Proc. Amer. Math. Soc. **112**(4), 959–964 (1991)
2. Anastassiou, G.A.: Bivariate left fractional polynomial monotone approximation. In: Intelligent Mathematics II: Applied Mathematics and Approximation Theory, pp. 1–13, AMAT, Ankara, May 2015. Springer, New York (2016)
3. Anastassiou, G.A.: Bivariate right fractional polynomial monotone approximation. In: Computational Analysis, pp. 19–31, AMAT, Ankara, May 2015. Springer, New York (2016)
4. Anastassiou, G.A.: Bivariate abstract fractional monotone constrained approximation by polynomials (2021) (submitted)
5. Anastassiou, G.A., Shisha, O.: Monotone approximation with linear differential operators. J. Approx. Theory **44**, 391–393 (1985)
6. Badea, I., Badea, C.: On the order of simultaneously approximation of bivariate functions by Bernstein operators. Anal. Numé r. Théor. Approx. **16**, 11–17 (1987)
7. Mamatov, T., Samko, S.: Mixed fractional integration operators in mixed weighted Hölder spaces. Fract. Calculus Appl. Anal. **13**(3), 245–259 (2010)
8. Shisha, O.: Monotone approximation. Pac. J. Math. **15**, 667–671 (1965)
9. Stancu, D.D.: Studii Si Cercetari Stiintifice**XI**(2), 221–233 (1960)

Conclusion

During the last 50 years fractional calculus due to its wide applications to many applied sciences has become a main trend in mathematics. Its predominant kind is the newer one of Caputo type with all of its variations because of their easier applicability. These involve usually singular kernels but recently also non singular kernels.

In this short monograph we employ an abstract kernel fractional calculus with applications to Prabhakar and non-singular kernel fractional calculi. Our results are univariate and bivariate.

In the univariate case we present abstract fractional monotone approximation by polynomials and splines, and in the bivariate case we give the abstract fractional monotone constrained approximation by bivariate pseudo-polynomials and polynomials. This monograph's results are expected to find applications in many areas of pure and applied mathematics, especially in fractional approximation theory and fractional differential equations. Other interesting applications can be in applied sciences like geophysics, physics, chemistry, economics and engineering, etc.

The advantage of this abstract approach is the wide range of its applications in covering the various natural phenomena.

G. A. Anastassiou, *Abstract Fractional Monotone Approximation, Theory and Applications*, Studies in Systems, Decision and Control 411, https://doi.org/10.1007/978-3-030-95943-2